Laser-assisted Formation of Metallic Nanoparticles

Theory, fabrication and applications

Online at: https://doi.org/10.1088/978-0-7503-6224-5

Laser-assisted Formation of Metallic Nanoparticles

Theory, fabrication and applications

Ritesh Sachan

School of Mechanical and Aerospace Engineering, Oklahoma State University, Stillwater, Oklahoma, USA

IOP Publishing, Bristol, UK

ISBN 978-0-7503-6224-5 (ebook)
ISBN 978-0-7503-6222-1 (print)
ISBN 978-0-7503-6225-2 (myPrint)
ISBN 978-0-7503-6223-8 (mobi)

DOI 10.1088/978-0-7503-6224-5

Version: 20251201

IOP ebooks

British Library Cataloguing-in-Publication Data: A catalogue record for this book is available from the British Library.

Published by IOP Publishing, wholly owned by The Institute of Physics, London

IOP Publishing, No.2 The Distillery, Glassfields, Avon Street, Bristol, BS2 0GR, UK

US Office: IOP Publishing, Inc., 190 North Independence Mall West, Suite 601, Philadelphia, PA 19106, USA

I offer my deepest gratitude to my family, ***Ritika****, my wife, and* ***Avyaan****, my son, for their unwavering love, patience, and encouragement. Their support has been a constant source of strength throughout the many stages of this work. I also extend my heartfelt thanks to the mentors, students, and collaborators who have, directly or indirectly, contributed to the knowledge, ideas, and inspiration behind this book. Their guidance, curiosity, and shared passion for research have shaped not only the scientific content presented here but also the journey that made it possible.*

Contents

Preface

The field of laser-assisted nanoparticle synthesis has grown from a niche experimental technique to a powerful and versatile platform for creating metallic nanostructures with unprecedented control. Over the past two decades, advances in ultrafast and nanosecond laser technologies, thin-film engineering, and in-situ diagnostics have transformed our ability to tailor matter at the nanoscale. Today, laser-driven fabrication routes not only offer rapid, clean, and highly tunable pathways for generating metallic nanoparticles, but also open new avenues for understanding the fundamental physics that govern their formation, evolution, and functionality.

This book, *Laser-Assisted Formation of Metallic Nanoparticles: Theory, fabrication, and applications*, is written to serve both as a research-focused reference and as an educational foundation for students and scientists entering this exciting domain. While the existing literature is extensive, it often remains fragmented across separate communities; laser processing, thin-film science, materials synthesis, and nanophotonics. My goal is to bridge these fields by presenting a comprehensive, integrated view of how lasers interact with matter and how these interactions can be harnessed to create nanoparticles with tailored size, shape, composition, and phases.

The motivations behind this book are twofold. First, it provides a structured summary of the significant research conducted worldwide, including insights gained through my own laboratory's work on nanosecond laser-induced dewetting and high-entropy alloy nanoparticles. Second, it serves as a practical guide to the experimental and theoretical tools that underpin this field. From thermal modeling and plume dynamics to thin-film instabilities, diffusion mechanisms, and advanced electron microscopy, each chapter aims to connect fundamental principles with real experimental workflows, challenges, and best practices.

To maintain accessibility without compromising scientific rigor, I have adopted an engaging, narrative tone. Readers will find theoretical concepts explained with intuitive physical arguments, supported by case studies, experimental examples, and high-resolution microstructural data. The book is intentionally structured to benefit a diverse audience, graduate students seeking clear concepts, early-career researchers looking for methodology, and experienced scientists pursuing deeper mechanistic understanding or new application pathways.

Laser-assisted nanoparticle formation continues to evolve, driven by emerging needs in catalysis, energy conversion, sensing, medicine, and quantum materials. It is my hope that this book not only captures the state of the art but also inspires new ideas and innovations. Whether you are designing combinatorial thin-film libraries, exploring extreme nonequilibrium transformations, or simply developing a curiosity for how light can sculpt matter, I hope these chapters will provide clarity, guidance, and a foundation for future discovery.

I am grateful to the students, collaborators, and research partners whose contributions have shaped much of the research described here. Their curiosity, persistence, and creativity continue to push the boundaries of what laser-driven materials synthesis can achieve.

Finally, I offer this book as an invitation to learn, to question, and to explore the remarkable ways in which lasers can create order, complexity, and function at the smallest scales.

Acknowledgments

This book is the result of years of exploration, collaboration, and support from many individuals and institutions. I am deeply grateful to my mentors, mentees, students and collaborators who played a role, direct or indirect, in shaping the ideas, experiments, and perspectives presented here. I also acknowledge the support of the National Science Foundation (NSF) for this work under CAREER award (CMMI- 2237820).

Author biography

Ritesh Sachan

Dr Ritesh Sachan is an Associate Professor in the Department of Mechanical and Aerospace Engineering at Oklahoma State University. His research focuses on uncovering the evolution mechanisms and structural transformations in nanoscale high-entropy materials through the combined use of thin-film science, nanosecond laser processing, and advanced electron microscopy. A major highlight of Dr Sachan's recent work is the development of a nanosecond laser–driven approach for fabricating material libraries of high-entropy alloy nanoparticles, a rapidly emerging class of materials with transformative potential for energy and catalytic applications. Dr Sachan earned his PhD from the University of Tennessee, Knoxville, and previously served as an NRC Research Fellow at the U.S. Army Research Office and a Postdoctoral Researcher at Oak Ridge National Laboratory. He has received numerous honors, including the NSF CAREER Award, ASM-IIM Lectureship Award, National Research Council Fellowship from the National Academy of Sciences, and the TMS Young Leader Award. To date, he has authored nearly 100 journal publications and delivered numerous invited and conference talks, reflecting his substantial contributions to the fields of nanomaterials, thin-film science, and laser-driven materials synthesis.

Chapter 1

Introduction

The rapid advancement of nanoscience and nanotechnology over the past few decades has unlocked extraordinary capabilities to manipulate matter at the atomic and molecular scale. At the forefront of this revolution are metallic nanoparticles (MNPs), nanoscale entities composed of metals that exhibit unique physical and chemical properties distinct from their bulk counterparts. Their exceptional surface-to-volume ratio, tunable size and shape, and quantum-scale effects have made them indispensable in diverse domains such as plasmonics, sensing, catalysis, energy conversion, data storage, and biomedicine.

Metallic nanoparticles are not merely scaled-down versions of bulk materials; they exhibit novel behaviors resulting from confinement effects, discrete energy levels, and surface-dominated reactivity. For instance, noble metal nanoparticles such as gold and silver exhibit localized surface plasmon resonance, where collective oscillations of conduction electrons in response to electromagnetic radiation lead to strong light absorption and scattering. Similarly, magnetic nanoparticles demonstrate superparamagnetism, while bimetallic or alloyed nanoparticles offer synergistic catalytic performance that cannot be achieved by individual constituents alone.

Traditional synthesis approaches for MNPs, such as chemical reduction, sol-gel processes, and thermal decomposition, often require surfactants, stabilizers, and harsh reagents. While effective, these methods may introduce impurities or impose limitations on structural purity, scalability, or environmental sustainability. In contrast, laser-based synthesis methods present a compelling alternative. These techniques rely on the interaction of high-energy, short-duration laser pulses with matter to induce localized melting, vaporization, and plasma formation. By harnessing these extreme yet highly controllable conditions, laser-based methods enable the fabrication of high-purity nanoparticles with minimal contamination, often in a single-step, solvent-free, and scalable manner.

Two dominant pathways in this arena are pulsed laser-induced dewetting and laser ablation in liquids (LAL). In pulsed laser-induced dewetting, thin metallic films

doi:10.1088/978-0-7503-6224-5ch1

deposited on substrates undergo instabilities and break up into discrete nanoparticles upon laser irradiation, driven by surface energy minimization and melt flow dynamics. This method enables precise patterning and the creation of ordered nanoparticle arrays. On the other hand, LAL involves focusing laser pulses onto a bulk metal target immersed in a liquid medium, where rapid energy deposition causes explosive material ejection, cavitation, and subsequent nanoparticle formation through condensation and complex fluid dynamics. These two routes not only provide a high degree of tunability in nanoparticle synthesis but also open new avenues for forming bimetallic, high-entropy, and complex heterostructures.

Understanding the microstructural evolution of laser-synthesized nanoparticles is critical to tailoring their properties. The non-equilibrium conditions inherent to laser-material interaction lead to rapid cooling rates, supersaturation, and metastable phase formation. These conditions often produce fine crystallites, amorphous domains, and unique interfaces, which significantly influence optical, magnetic, thermal, and catalytic behavior. Advances in electron microscopy, spectroscopy, and atomistic simulations have greatly improved our ability to probe these structures at the atomic level, providing insights into nucleation, growth mechanisms, and interfacial phenomena.

This book is organized to provide a holistic understanding of laser-based nanoparticle synthesis and its implications for material design. Chapter 2 delves into pulsed laser-induced dewetting, exploring fluid mechanical models, atomistic simulations, and experimental realizations, while also extending to bimetallic nanostructures. Chapter 3 focuses on laser ablation in liquid, covering the fundamental physics, chemical interactions, and recent advancements in hybrid synthesis techniques and nanopatterning. Chapter 4 presents cutting-edge progress in synthesizing high-entropy nanoparticles, quantum dots, nanoparticle-polymer composites, and other functional nanostructures enabled by laser processing.

In this book, we further highlight the diverse functional properties of metallic nanoparticles, from their optical, magnetic, thermal, and electrical responses to their utility in green energy technologies, catalysis, and biomedical applications.

Altogether, this book aims to serve as a comprehensive reference for scientists, engineers, and students interested in the interdisciplinary science of laser-nanomaterials synthesis. By combining theoretical foundations, experimental techniques, and real-world applications, it offers a roadmap for exploring how laser-based methods can reshape the future of nanomaterial development and deployment.

Chapter 2

Pulsed laser-induced nanoparticle formation: self-organization, dewetting and thin films

In this chapter, we discuss the fundamentals of metallic nanoparticle formation using pulsed laser-induced dewetting (PLiD) of thin films and nucleation-growth phenomena. The interaction of ions, electrons, and photons with matter underpins both bottom-up and top-down approaches to the fabrication of nanostructures. These interactions enable energy transfer to the irradiated material, which can induce processes such as self-assembly or controlled material removal, leading to the formation of nanoscale features.

Among these methods, laser–matter interaction has gained prominence as a highly versatile and precise tool for structuring materials at the nanoscale. This book specifically focuses on recent advancements in the use of laser-based techniques for nanofabrication. The development of fast and ultrafast pulsed lasers—operating in the nanosecond, picosecond, and femtosecond regimes—has significantly expanded the potential of laser-assisted processing. These lasers enable the fabrication of a wide variety of nanostructures by interacting with thin films or bulk materials deposited on functional substrates. The process allows for refined control over the resulting nanostructures' shape, size, crystallinity, and composition through careful adjustment of laser parameters such as pulse duration, fluence, and repetition rate.

One of the primary advantages of laser-based nanofabrication lies in its ability to manipulate materials across a broad size range, from the micrometer scale down to the nanometer regime. The short pulse durations enable ultrafast energy deposition and rapid quenching, thereby minimizing thermal damage to the surrounding substrate and adjacent regions. Based on the underlying mechanism, this process for nanoparticle formation is generally known as pulsed laser-induced dewetting of thin films that transforms an energetically unstable metallic film into metallic nanoparticles. In addition, the non-contact nature of laser processing makes it suitable for fabricating on non-planar surfaces. This approach is also compatible

doi:10.1088/978-0-7503-6224-5ch2

with various post-processing steps, including surface chemical modifications, allowing for greater flexibility and integration into broader fabrication workflows.

Taken together, these capabilities make laser–matter interaction a powerful technique for producing complex nanostructured materials with high spatial, temporal, and structural control, offering valuable opportunities for applications in electronics, photonics, and energy technologies.

2.1 Pulsed laser-induced dewetting

Dewetting is a Nature-inspired concept to create self-organized nano/microstructures. In Nature, water droplet formation on glass windows is a typical example of this phenomenon, where a thin film of water spontaneously breaks up into droplets on the glass surface due to overpowering interfacial surface energies. In the case of pulsed laser-induced dewetting, irradiation from a high-energy ultraviolet laser on ultrathin metallic films (1–30 nm) allows its homogeneous melting within a few nanoseconds [1]. This melt-phase thin film becomes unstable under two competing energy terms. In the cases of most metallic films studied, these two energies correspond to the stabilizing surface tensions and the destabilizing attractive intermolecular dispersion forces between the film-substrate and film-vacuum interfaces. If the attractive forces dominate, films spontaneously rupture, eventually forming nanostructure morphologies with well-defined length scales via nanoscale mass transport [1–5] (figure 2.1).

Generally, there are several physical properties related to (i) the thin films (ii) substrates (thickness, surface energies, dielectric function governing reflectivity and absorptivity, melting and vaporization temperatures, thermal conductivity, electron-specific heat, lattice heat capacity, and electron–phonon coupling) and (iii) irradiated lasers (pulse duration, energy, wavelength, repetition rate) governing the overall system free energy leading to the thin film dewetting and making the nanoparticle formation possible [6]. However, in the dewetting scenarios of thin films on insulating substrates with an ultraviolet nanosecond pulsed laser (KrF, 248 nm wavelength, 25 ns pulse width), the nanoparticle size and spacing vary as a function of the film thickness depending on the thickness effect on the energy terms [7–9]. The formation of nanoparticles over a broad range of achievable sizes has been extensively investigated by controlling the parameters in single metals (e.g. Ag,

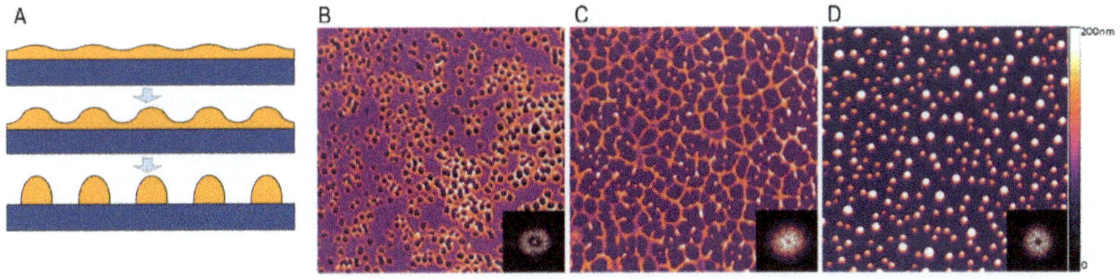

Figure 2.1. (A) Schematic and (B–D) SEM images showing the progression forming nanoparticles via pulsed laser-induced dewetting of thin films. Reproduced from [2] with permission from the Royal Society of Chemistry.

Au, Co, Ni, Fe, Cu, Pt) [8, 10–15], semimetals (Si) [16, 17] and bimetallic (e.g. AgCo, AgNi, AuCo, CuNi, CuCo) systems [18–26].

2.2 Factors influencing laser irradiation

The outcome of laser-induced structuring of metal films is strongly influenced by the intrinsic properties of the metal film, the nature of the substrate, and the characteristics of the laser used. Depending on these parameters, a diverse array of nanostructures and surface features can be produced. These include metal nanoparticles, which have been widely reported and studied across various materials and laser conditions, as well as metal microbumps formed through localized melting and resolidification processes. The formation mechanisms, morphology, and functional properties of these structures have remained subjects of extensive investigation, with significant research efforts continuing up to the present day [8, 27–29]. This ongoing interest reflects the importance of laser–matter interaction as a tool not only for nanoparticle synthesis but also for the fabrication of complex and ordered surface topographies relevant to applications in plasmonics, optoelectronics, sensing, and surface-enhanced spectroscopies.

When the laser is incident on the surface of metallic thin film, the interaction between the laser and the thin film depends on the nature of the properties of the metallic film, the substrate and the laser properties. Figure 2.2 summarizes various factors affecting the nanostructurization of thin films [6]. As shown in the figure, the

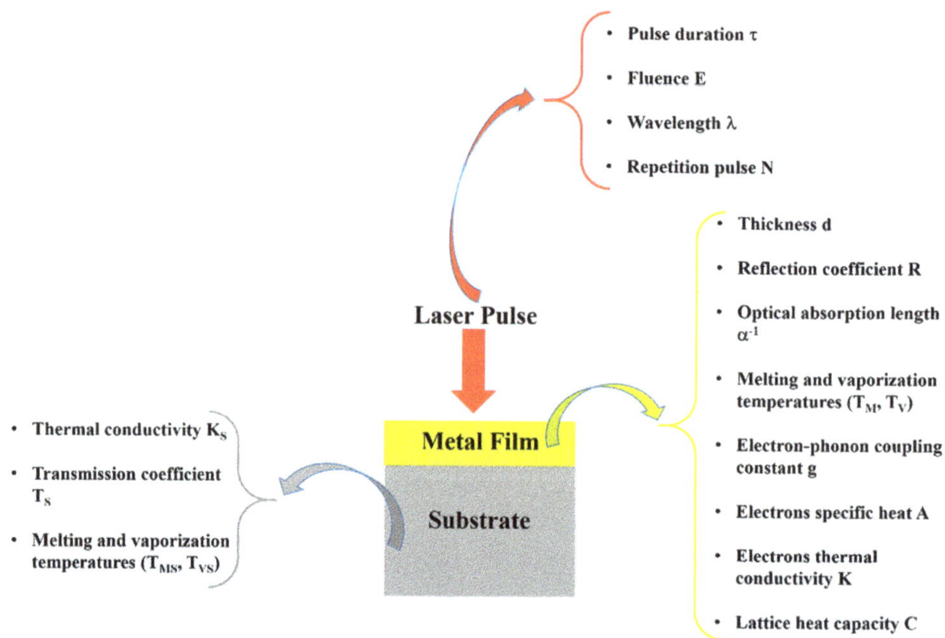

Figure 2.2. The schematic representation of a metal film deposited on a substrate and processed by a laser pulse to induce the film nanostructuration. In the figure, some critical parameters concerning the laser, the film, and the substrate affecting the nanostructuration process are indicated. Reproduced from [6]. CC BY 4.0.

primary characteristics of the laser that affect the nanostructure formation are the pulsed duration, fluence, laser wavelength and repletion rate. A critical aspect among these is the pulsed duration that can vary from femtoseconds to nanoseconds and leads to the formation of a nanostructure with different morphologies [8, 30–33]. Similarly, the physical properties, such as thermal conductivity, optical properties and melting/boiling points, play a defining role in the melting and solidification process under laser irradiation. The properties of the substrate control the heat dissipation at the film-substrate interface and regulate the solidification process to create nanostructures.

2.3 Liquid–film interactions

The fundamental step in utilizing lasers as a nanofabrication tool is to understand how laser energy interacts with matter and how this interaction translates into heating, melting, and subsequent structural modifications at the nanoscale. When a pulsed laser irradiates a solid, the incoming photons are absorbed and act as an internal energy source. The material's response is governed by the complex interplay of electronic excitations, electron–electron interactions, and electron–phonon coupling, all of which occur on ultrafast timescales [6, 34].

For photon energies in the near-infrared to near-ultraviolet range, absorption occurs primarily through electronic excitations. In semiconductors and insulators, photons promote electrons from the valence band to the conduction band, creating electron–hole pairs. The subsequent recombination of these pairs restores equilibrium on the nanosecond timescale, which is strongly dependent on the material's band structure [6]. In contrast, in metals, laser absorption is dominated by intraband transition scattering events where laser photons are absorbed by the free electrons of the metal, leading to nonequilibrium electron distributions within just a few femtoseconds [34, 35]. A preliminary approach to describing metal heating under laser irradiation can be based on the Drude model. Within this model, the electron scattering mean time is connected to the free electron gas conductivity by $\sigma = \mathrm{ne}^2\tau_D\,\mathrm{m}^{-1}$, with σ being the conductivity, n being the electron density, m being the electron mass, and τ_D being the electron scattering mean time, typically a few femtoseconds.

Immediately following absorption, the excited electron population undergoes ultrafast thermalization through electron–electron scattering. This process of electron–electron thermalization, occurring on the order of 1–10 femtoseconds, redistributes the absorbed energy among the electron ensemble, resulting in a Fermi–Dirac-like distribution characterized by an elevated electron temperature (T_{ee}). At this stage, the lattice remains relatively unaffected.

The next step involves electron–phonon coupling, where the hot electron gas transfers energy to the atomic lattice. This occurs on the 0.1–1 picosecond timescale, and is significantly slower than electron–electron scattering. The characteristic electron–phonon relaxation time (τ_{ep}) depends on the material's electron density, band structure, Debye frequency, and electron–phonon coupling constant (G_{ep}). Transition metals such as Cr, Mo, W, and Fe exhibit strong electron–phonon coupling (large G_{ep}), which enables rapid heating of the lattice and efficient ablation under intense laser irradiation. In contrast, noble metals such as Au and Ag display weaker coupling (smaller G_{ep} values), resulting in slower lattice heating and the persistence of a molten phase for tens of picoseconds [35, 36]. For example, in Au, the electron subsystem transfers its energy to the lattice within ~15 ps, and complete equilibrium between electrons and phonons is typically reached within ~50 ps [35, 37].

This dynamic separation of timescales is elegantly described by the two-temperature model (TTM). In this framework, the electronic and lattice subsystems are treated as coupled but distinct heat reservoirs. The evolution of T_e and T_L is governed by coupled differential equations: the electron equation accounts for laser absorption, diffusion, and electron–phonon energy transfer, while the lattice equation tracks the subsequent heating driven by electron energy relaxation [38, 39]. The TTM not only explains the ultrafast nonequilibrium conditions in metals but also provides predictive capability for understanding phase transitions, ablation thresholds, and nanoparticle formation during pulsed-laser processing.

In summary, laser–matter interaction in metals unfolds across three distinct but interconnected stages:

1. **Femtosecond electron excitation and thermalization**—photon absorption and rapid electron–electron scattering establish a hot electron gas.
2. **Picosecond electron–phonon coupling**—energy is transferred from electrons to the lattice, raising the lattice temperature.
3. **Nanosecond lattice response**—melting, resolidification, ablation, and nanostructure formation occur as the lattice equilibrates.

The relative timescales of these processes vary with material properties such as electron density, band structure, and coupling constants. This hierarchy of ultrafast events is central to the physics of laser nanofabrication, determining not only whether a material melts, ablates, or restructures, but also the morphology and microstructure of the resulting nanoparticles and thin films.

2.4 Nanoscale transport and dewetting

The classical dewetting instability in thin films can be interpreted as a competition between two energy terms. In the case of a large number of polymer or metallic films studied, these two energies correspond to the surface tension and the attractive intermolecular dispersion force between the film–substrate and film–vacuum interfaces mediated by the film material.

At the nanoscale, the interaction between a molten metallic film and its supporting substrate is a critical factor governing the stability of ultrathin films. These interactions are generally expressed through the concept of disjoining pressure (Π), which depends on the film thickness, h, and arises from the competition between long-range attractive and short-range repulsive forces. In metallic liquids, Π reflects not only the van der Waals interactions between atoms and molecules but also contributions from electronic interactions, such as repulsion between conduction electrons. This pressure term, therefore, provides a unified description of the net stabilizing or destabilizing influence acting on a thin liquid film [40, 41].

In its simplest approximation, liquid metal–substrate interactions can be described using London dispersion forces, where instantaneous dipole fluctuations in neutral atoms induce dipoles in neighboring atoms, giving rise to pairwise forces that scale as r^{-6} with interatomic separation [42]. A more refined treatment invokes macroscopic van der Waals forces, from which mean-field expressions for the disjoining pressure can be derived. While systematic comparisons between theory and experiment remain limited, especially for metallic films where direct measurements are challenging, various functional forms have been proposed based on analogy to polymeric liquids, where such studies are easier to control [23, 43, 44].

The most widely used general expression for disjoining pressure is written as [45, 46]:

$$\Pi\,(h) = \kappa\, f\,(h) = \kappa \left[\left(\frac{h_*}{h}\right)^n - \left(\frac{h_*}{h}\right)^m \right]$$

where $n > m$ are positive exponents, h_*^3 is the equilibrium film thickness at which attractive and repulsive forces balance, and κ is a pressure scale related to the Hamaker constant by $A = 6\pi\kappa h_*^3$. In most modeling efforts, only a limited range of exponent pairs $(n,\ m)$ has been utilized, and for metallic liquids, reduced forms containing solely attractive contributions are often adopted [10, 47, 48]. More refined approaches have incorporated exponential terms to account for short-range forces and substrate-dependent interactions [8].

Although no single expression for Π in liquid metals has been universally established, the framework remains highly effective in linking molecular-level interactions to mesoscale processes such as film rupture, spinodal instabilities, and nanoparticle patterning during laser exposure. The interplay of attractive and repulsive contributions within the disjoining pressure provides an essential theoretical basis for understanding the morphological dynamics of molten films.

In thin-film hydrodynamics, the role of disjoining pressure is incorporated through a modified lubrication approximation, which describes the temporal evolution of the film thickness $h(x,\ y,\ t)$. The governing equation for a viscous molten film of viscosity μ under capillary and substrate interaction forces is:

$$\frac{\partial h}{\partial t} = -\,\nabla \cdot \left[\frac{h^3}{3\mu}\, \nabla\, (\gamma\, \nabla^2\, h + \Pi(h) \right]$$

where γ is the surface tension and $\Pi(h)$ is the disjoining pressure. The first term in the bracket reflects curvature-driven transport governed by capillary forces, whereas the second term represents stabilizing or destabilizing influences arising from long- and short-range intermolecular interactions. The competition between these contributions dictates whether the molten film remains stable, ruptures into distinct patterns, or undergoes spinodal dewetting that produces ordered nanoparticle arrays.

For metallic films of nanometer-scale thickness exposed to pulsed-laser heating, the model predicts that film stability is determined by the initial thickness. Sub-equilibrium films ($h < h_*$) are destabilized by long-range van der Waals forces, whereas films exceeding h_* may persist in a metastable state, only breaking apart when perturbed by thermal or structural irregularities. Introducing $\Pi(h)$ into the thin-film framework provides a rigorous connection between molecular-scale forces and the macroscopic dewetting patterns leading to nanoparticle formation.

The dewetting mechanism can be understood by considering the energy transfer and/or energy conversion in systems during liquid flow. The dewetting behavior under nanosecond laser heating can be understood in terms of maximizing the rate of energy transfer between the thermodynamic state of the film and its viscous flow on the substrate. For an ultrathin metal film on an inert substrate such as glass (SiO_2), the total free energy of the system is made up of two terms: the surface/interface energies γf_v and γf_s of the film-vacuum and film-substrate interfaces, respectively, and the disjoining pressure $\Pi(h_o)$, where h_o is the initial height of the film. The disjoining pressure $\Pi(h_o)$ is typically comprised of an attractive dispersive interaction between the metal film-substrate and film-vacuum interface of type $A/12\pi h_o^2$, where A is the Hamaker coefficient with a negative sign, and a short-range repulsive term between the film and substrate atoms, typically given by B/h_o^n, where n is normally $\geqslant 4$. For ultra-thin films, the gravitational energy term is negligible, as is the short-range repulsive term, and will therefore be ignored in the subsequent analysis.

For such thin films, a height perturbation of the liquid film surface (figure 2.3), such as by capillary waves, results in a change in free energy. The contributions to this free energy change come from two terms. First, the area of the film surface increases with respect to its original flat state for perturbations of any wavelength, and hence the surface tension will contribute a positive term to the free energy change. Second, the attractive dispersion energy (which is proportional to $\Pi(h_o)^n \propto 1/h_o^4$) contributes a large negative term. Therefore, the resulting total free energy change can be negative when the wavelength of the perturbation is of appropriate value, thus implying conditions for a spontaneous break-up of the liquid film [10]. In the case of spontaneous dewetting, the characteristic patterning length scale λ is known to scale with the initial film thickness as $\lambda \propto h_o^2$ [8, 10, 20]. Dewetting of metallic liquid films of

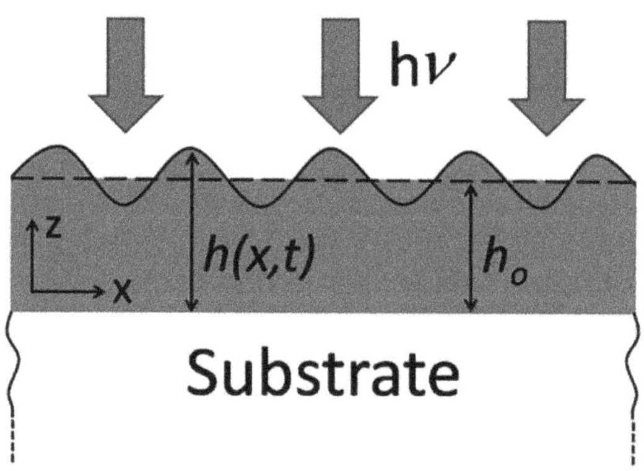

Figure 2.3. Schematic illustrating laser heating of ultrathin liquid films. A liquid metal film of initial height h_o undergoes surface height fluctuations due to capillary waves. This non-uniform height couples to the incident laser light and results in a local height-dependent temperature. Reproduced from [49], with permission from Springer Nature.

nanometer thickness under nanosecond laser melting shows this classical behavior. These results can be understood in terms of the principle of maximizing the rate of energy transfer between the thermodynamic state of the film and its viscous flow on the substrate.

2.5 Experimental results on metallic film dewetting

The concept of using dewetting to synthesize metallic nanoparticles, while widely explored over the past 25 years, traces its origins to studies of polymer thin films that span more than five decades. In polymeric systems, dewetting has long been recognized as a powerful route to understand interfacial instabilities and pattern formation at the nanoscale. The most significant finding is that thermodynamically unstable polymer-substrate films undergo spontaneous dewetting, producing well-defined intermediate and final morphologies governed by characteristic length scales. Importantly, the free energy landscape of such thin films as a function of thickness exhibits strong analogies to the composition-dependent free energy in binary alloys that undergo spinodal decomposition. This resemblance has led to the description of the process as spinodal dewetting, where amplification of thickness fluctuations in the unstable regime drives self-organized rupture of the film and the eventual emergence of periodic structures.

When adapted to metallic thin films, this framework provides a direct pathway to nanoparticle formation under thermal or laser-driven melting. The parallels to spinodal decomposition emphasize the universal character of the instability mechanism: in both polymer and metal films, free energy curvature dictates whether small perturbations are suppressed or amplified, thereby determining whether the film remains stable, dewets into droplets, or organizes into periodic nanostructures. This analogy underscores why dewetting is not only a fundamental physical phenomenon but also a versatile nanomanufacturing strategy, enabling controlled fabrication of ordered nanoparticle arrays through self-organization.

For ultrathin films in the thickness range of approximately 1–100 nm, gravitational effects can be neglected, and the stability of the film is instead governed by intermolecular and interfacial forces. In this regime, the curvature of the free energy with respect to film thickness is negative (i.e., $d^2G/dh^2 < 0$), as illustrated in figure 2.4. Such a condition defines a thermodynamically unstable state, closely resembling the free energy curvature characteristic of the spinodal regime in binary alloys undergoing phase segregation. This analogy has provided a strong conceptual framework for understanding thin-film dewetting as a nanoscale counterpart of spinodal decomposition.

A second critical observation from dewetting studies of unstable films is the progression of morphological transformations from an initially smooth liquid film to the final array of droplets. This transition proceeds through intermediate states with well-defined length scales that arise spontaneously due to the amplification of surface perturbations. Two principal pathways of film breakup are commonly observed. In the first, rupture begins with the nucleation of holes that expand and coalesce, giving rise to polygonal cellular structures before breaking up into droplets.

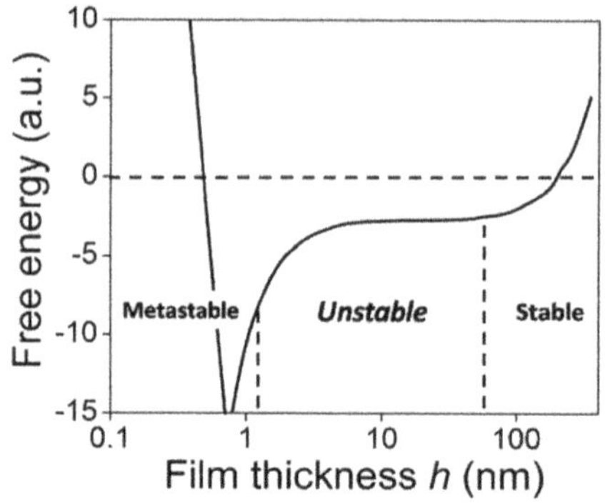

Figure 2.4. The free energy curve of a metallic film. There exist three distinct stability regimes in the film, including an unstable, a metastable, and a stable thickness regime. Typically, metal films are unstable from 0 to 100 nm, while films > 100 nm are stable. Reproduced from [50], with permission from Springer Nature.

In the second, the film destabilizes into bicontinuous structures, where labyrinthine networks of thinning film evolve before fragmenting into isolated particles. Both pathways are characteristic of the instability regime associated with film thicknesses between 1 and 100 nm, where dewetting occurs spontaneously and leaves behind self-organized morphologies with distinct periodicity.

These intermediate morphologies—holes, polygonal domains, and bicontinuous networks—are not just transient states but carry information about the underlying instability mechanism, characteristic length scales, and energy minimization pathways. Their universality across polymeric, metallic, and even oxide films highlights the fundamental role of interfacial thermodynamics in driving self-organization at the nanoscale.

The formation of nanoparticles over a broad range of achievable sizes has been extensively investigated by controlling the parameters in single metals (e.g. Ag, Au, Co, Ni, Fe, Cu, Pt) [8, 10–15], semimetals (Si) [16, 17] and bimetallic (e.g. AgCo, AgNi, AuCo, CuNi, CuCo) systems [18–26]. Here, the nanoparticle formation results from the surface perturbations caused by intermolecular forces, specifically in the thin films during laser irradiation, followed by the surface tension that governs the final nanoparticle shape. Nanosecond laser-induced dewetting is also distinguished from the laser ablation process, which relies on collecting ablated nanoparticles in liquids via rapid quenching of the plasma form during direct laser irradiation on bulk materials. The driving force behind the formation of the nanoparticles is the breakup of energetically unstable thin films under laser irradiation, leading to the accumulation of material in a nanoscale droplet shape within nanoseconds. Figure 2.5 shows the formation of Fe nanoparticles obtained from the dewetting of 3.5 nm thin film on a SiO_2 substrate. As shown in the figure, the film spontaneously ruptures, as demonstrated in the formation of holes and finally transforms into droplet-shaped nanoparticles during subsequent irradiation due to the accumulation of liquid metal.

It has been seen that the nanomorphology resulting in nanoparticle formation via dewetting of a spinodal-like system shows a dependence on the film thickness h.

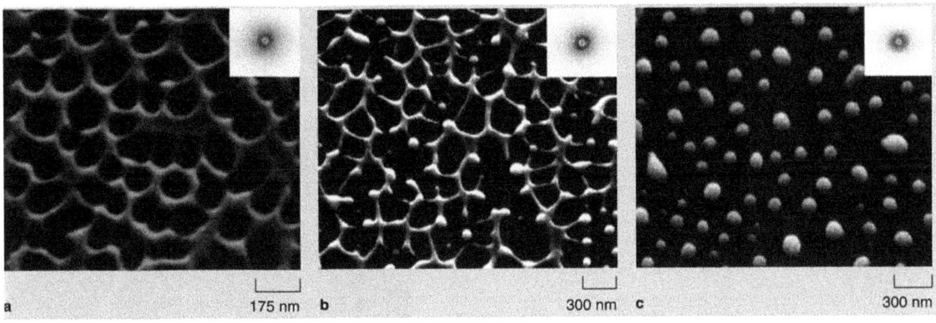

Figure 2.5. Scanning electron microscopy images depicting the characteristic stages of the morphological evolution of a dewetting ~3.5 nm iron film under ns pulsed laser irradiation. (a) The initial stage of the formation of holes in the film after a number of pulses $n \sim 5$. (b) An intermediate stage consisting of a polygon network of iron after $n \sim 500$. (c) The final particulate state with the nanoparticles forming at the vertices of the polygons after $n \sim 10\,000$. The fast Fourier transform in the inset of each of the morphological stages in images (a–c) depicts the short-range spatial order present during each stage of dewetting. Reproduced from [50], with permission from Springer Nature.

From an energetic perspective, the instability of thin films can be understood by evaluating changes in the system's free energy when perturbations are introduced to the film height. Such analyses predict that, for certain perturbation wave vectors, the film becomes unstable and can spontaneously dewet. The instability grows along the fastest-amplifying wave vector, which sets the characteristic length scale as a function of film thickness. Consequently, nanoparticle arrays form with thickness-dependent spacing. In general, the characteristic length (Λ) for spinodal-type dewetting exhibits a direct scaling with the initial film thickness.

$$\Lambda(h) = \sqrt{\frac{16\pi^3\gamma}{A}}h^2$$

where γ is the surface tension of the metal and A is the Hamaker coefficient for the vacuum–metal–substrate system. Figure 2.6 demonstrates an example of the thickness-dependent formation of a Ag nanoparticle where the film thickness varies between ~1 and 20 nm. The result of measuring length scales from these progressions is shown in figure 2.6(a). The early stage behavior is shown by the closed squares, the intermediate stage is shown by open triangles, while the final nanoparticle state is shown by open circles. One important observation from this measurement is that no dramatic change in length scale is seen when the morphology changes from the bicontinuous to the hole structures, i.e. between 9 and 11 nm. This strengthens the argument that both morphologies can arise for spinodal dewetting of Ag. Since the characteristic length scale for spinodal dewetting is known to vary as $\lambda \propto h^2$, we have also plotted h^2 trend lines for the early stage (solid line) and nanoparticle stage (dotted line) data sets with a h^2 trend. It is also seen that the nanoparticle formation takes different pathways during the depending process under the film instability, as shown in figure 2.7. However, other studies have reported that h^2 dependence can vary to $h^{3/2}$ depending on the material's properties [1, 22].

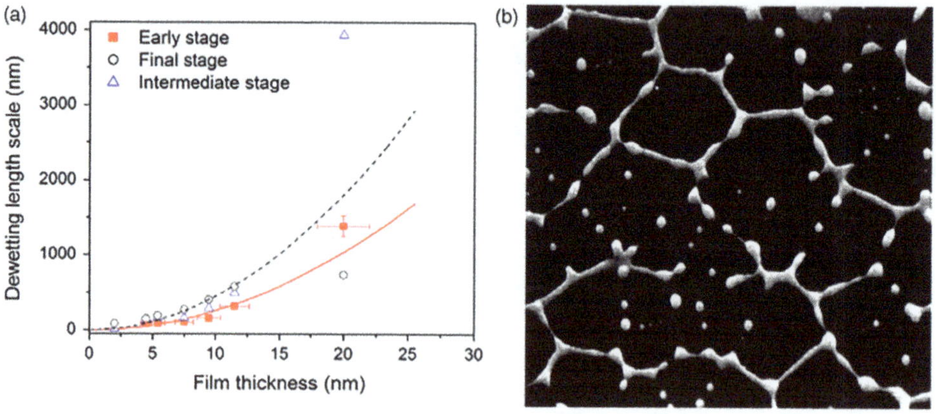

Figure 2.6. (a) Plot of the characteristic length scale for various stages of progression as a function of film thickness (h). The early stage behavior is shown by the closed squares, the intermediate stage is shown by open triangles, and the final nanoparticle state is shown by open circles. Trend lines with h^2 variation for the early stage (solid line) and nanoparticle stage (dotted line) data are also shown. For clarity, the error bars for the intermediate and final states are not shown. (b) SEM image ($15 \times 15\ \mu m^2$) of the intermediate stage to nanoparticle stage transition for a 20 nm film showing that the break-up of the arms of the polygons is via a Rayleigh-type process leading to multiple nanoparticles in each arm. Reproduced from [8].

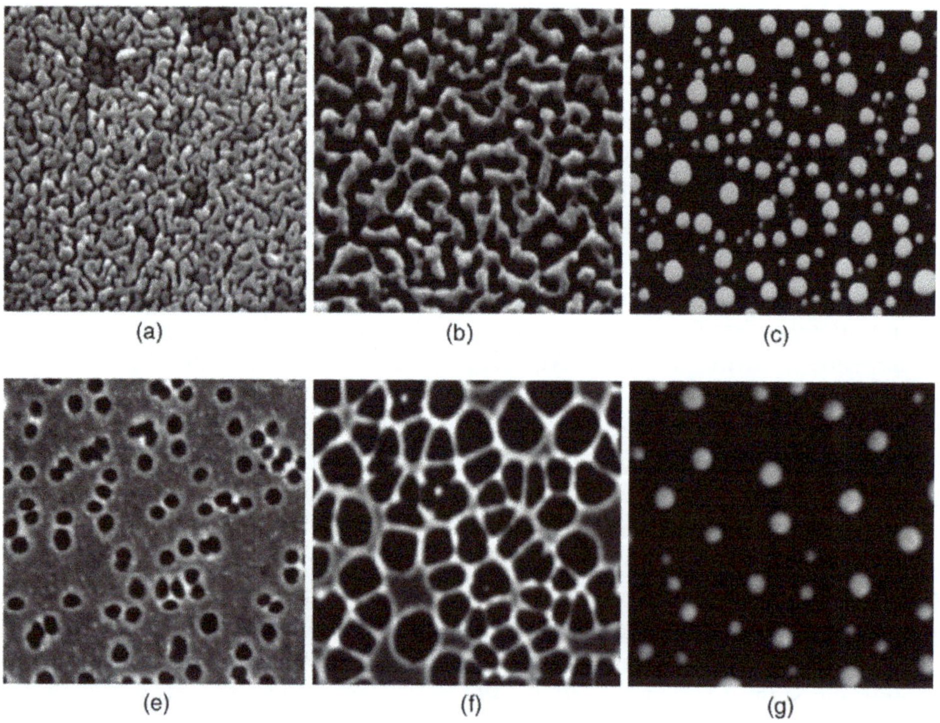

Figure 2.7. (a–c) SEM images ($1.5 \times 1.5\ \mu m^2$) of the progression of dewetting in the 4.5 nm thick film with increasing number of laser pulses (10, 100, 10 500). (e–g) Progression ($3 \times 3\ \mu m^2$) of dewetting in the 11.5 nm thick film with increasing number of laser pulses (10, 100, 10 500). Reproduced from [8].

In another interesting observation, it has been seen that there is a transition in the formed intermediate stages from bicontinuous to hole morphology as the thickness of the film increases. Figure 2.7 shows this morphological shift in Ag thin films, where figure 2.7(a–c) demonstrates the dewetting of a 4.5 nm Ag film following bicontinuous shaped intermediate stages, whereas the 11.5 nm thick film ruptures following hole-like morphology. This transition in the intermediate stages is correlated to the nature of the free energy of thin films. Specifically, the location of minima in the free energy curvature governs this transition thickness. The sum of different types of attractive and repulsive intermolecular interactions determines the total free energy and its curvature as a function of thickness. It has been seen that the repulsive forces dominate in the thinner films and as the film thickness increases, attractive forces overcome, thus making an energy minimum where the morphological transition occurs. The value of energy minima changes with material properties and has been calculated for different metals in previously reported work by Krishna *et al* [8].

2.6 Bimetallic film dewetting

Over the last decade, the application of the PLiD process has been exhaustively extended beyond the monometallic nanoparticles to bimetallic nanoparticles. This includes investigations on various bimetallic (e.g. AgCo, AgNi, AuCo, CuNi, CuCo) systems [18–26]. The major difference in the fundamental mechanism of the dewetting of bilayers to form bimetallic nanoparticles was explained by Krishna *et al* [27]. which suggested that the different bilayer arrangements change the signs of intermolecular interactions. This, in return, changes the mode of coupled deformations and the patterning characteristics.

As shown in figure 2.8(a), the initially flat film thicknesses for the bottom layer and the total layer are indicated as h_{10} and h_{20}, respectively, in the bilayer arrangement. The competing forces in this dewetting process were considered to be the surface or interfacial tensions and the long-range intermolecular dispersion forces between the various interfaces [27]. The surface tensions corresponding to the liquid$_1$–liquid$_2$ and liquid$_2$–gas interfaces are γ_{12} and γ_2, respectively. The three relevant dispersive forces, expressed in sign and magnitude by the Hamaker

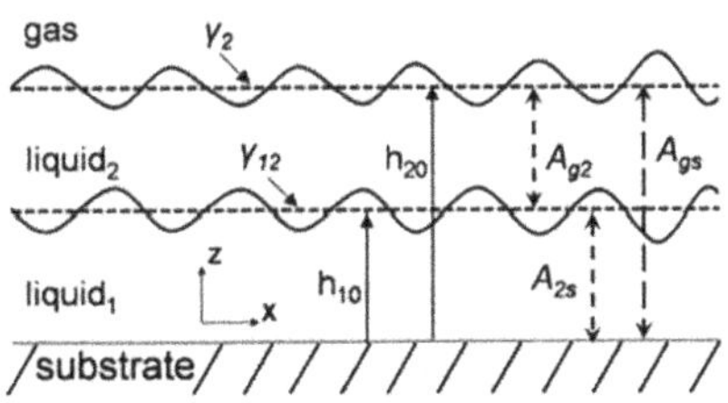

(a)

Hamaker Coeff.	Ag/Co/SiO$_2$	Co/Ag/SiO$_2$
A_{g2}	+	-
A_{gs}	-	-
A_{2s}	-	+

(b)

Figure 2.8. (a) Schematic of the bilayer arrangement. The perturbation depicted by solid surfaces is a squeezing mode. (b) Table depicting the signs of the Hamaker coefficients for the various pairs of interfaces in (*a*). A negative sign corresponds to a destabilizing attractive interaction. Reprinted with permission from [27]. Copyright (2011) American Chemical Society.

coefficient, are Ag2, Ags, and A2s, which correspond to interactions between the top-middle, top-substrate, and middle-substrate interfaces, respectively, as depicted in figure 2.8(a). The signs of these interactions for the two bilayer systems studied are tabulated in figure 2.8(b). The subscripts *g*, 2, and s denote the gas–liquid 2 (top layer), liquid 2–liquid 1 (bottom layer), and liquid 1–substrate interfaces, respectively. As can be seen, the perturbations occur at various existing interfaces in this bilayer arrangement compared to the monolayer arrangement and thus differ in the length scales of effective film rupture and eventually nanoparticle formation. An example of bilayer Ag/Co nanoparticle formation is demonstrated in figure 2.9.

Figure 2.9(a) shows the scanning electron micrograph of a typical nanoparticle array, synthesized from the dewetting of the Co (5 nm)/Ag (5 nm)/SiO_2 bilayer system. The inset shows the particle size histogram, typifying the 'narrow size distribution' achieved by this process. Also shown in the inset is the fast Fourier transform (FFT) of the contrast in the image. The annular ring evident in the FFT is

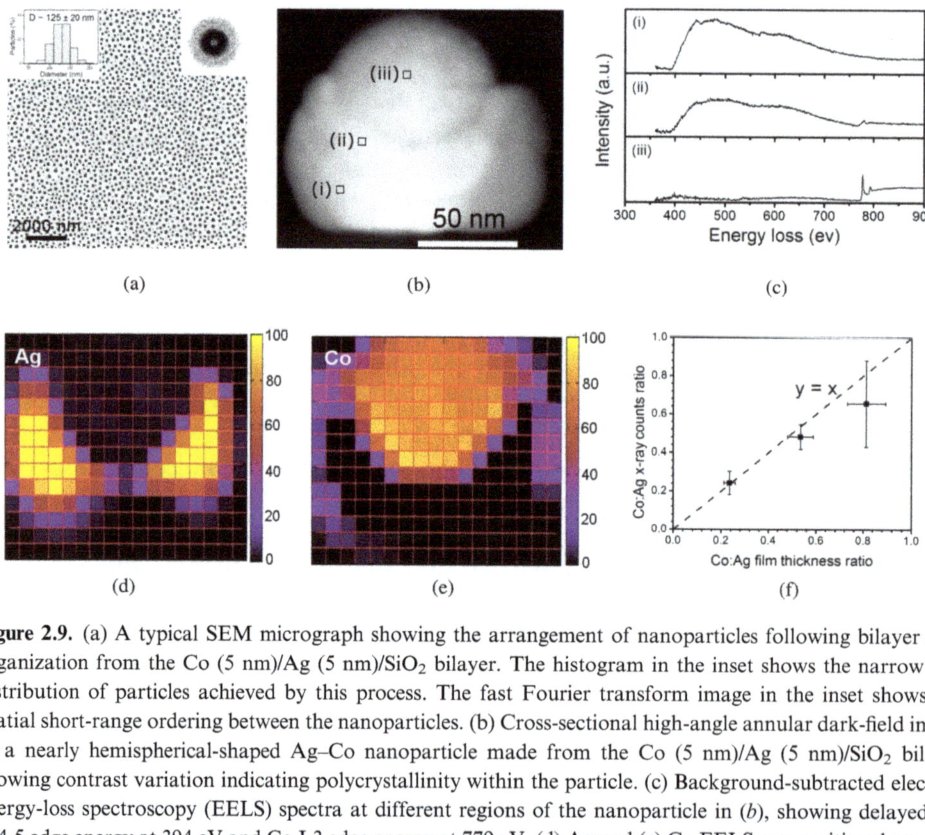

Figure 2.9. (a) A typical SEM micrograph showing the arrangement of nanoparticles following bilayer self-organization from the Co (5 nm)/Ag (5 nm)/SiO_2 bilayer. The histogram in the inset shows the narrow size distribution of particles achieved by this process. The fast Fourier transform image in the inset shows the spatial short-range ordering between the nanoparticles. (b) Cross-sectional high-angle annular dark-field image of a nearly hemispherical-shaped Ag–Co nanoparticle made from the Co (5 nm)/Ag (5 nm)/SiO_2 bilayer showing contrast variation indicating polycrystallinity within the particle. (c) Background-subtracted electron energy-loss spectroscopy (EELS) spectra at different regions of the nanoparticle in (*b*), showing delayed Ag M4,5 edge energy at 394 eV and Co L3 edge energy at 779 eV. (d) Ag and (e) Co EELS compositional maps of the enclosed region shown in image (b), exhibiting the immiscibility of Co and Ag in each other. The step size of the compositional map is 5.9 nm × 5.9 nm. The contrast bar shows the variation of atom % of Ag and Co individually in different locations of the enclosed region of image (*b*). (f) The average Co:Ag ratio in each nanoparticle from x-ray mapping (symbols) plotted against the Co:Ag film thickness ratio. Reproduced from [19].

indicative of a spatial short-range order in the nearest-neighbor spacing between the nanoparticles, consistent with the dewetting self-organizing process. From such SEM images, the average diameter of the particles in an array was established. The structure and composition of the individual nanoparticles were analyzed using transmission electron microscopy and electron energy-loss spectroscopy (EELS) measurements. A representative cross-sectional high-angle annular dark-field image of a ~100 nm nanoparticle obtained from dewetting of the Co (5 nm)/Ag (5 nm) bilayer on a SiO_2 substrate is shown in figure 2.9(b) [19].

As is evident, the nanoparticle contains several grains. The representative background-subtracted EELS spectra corresponding to different regions in figure 2.9(b) are shown in figure 2.1(c). EELS spectra show the presence of (i) pure Ag, (ii) the co-existence of Ag and Co and (iii) pure Co in different regions, respectively. The spectra showed a delayed Ag M4,5 edge at 394 eV and Co L3 edge at 779 eV. The spectrum from the co-existence regions of Ag and Co (figure 2.9(c)–(ii)) showed no shift in Ag and Co edge energies with respect to pure Ag and Co, respectively. This indicated that Ag and Co are clearly phase-segregated within the nanoparticles, as expected from the equilibrium thermodynamic phase diagram, and the mixed spectrum originated from the overlapping of Ag and Co grains along the electron beam path at this location [19]. Figure 2.9(d) and (e) show the elemental distribution of Ag and Co, respectively, inside the particle, analyzed by EELS near the M4,5 edge of Ag and the L3 edge of Co. The contrast bars in figures 2.1(d) and (e) represent the atomic % of Ag and Co at different regions (square blocks) of the nanoparticle. From this analysis, it is once again clear that Ag and Co are spatially segregated within the nanoparticle. The EELS analysis confirmed a negligible presence of oxygen in the Ag-rich region. In addition to the immiscible systems, such as Ag–Co and Ag–Ni, thermodynamically miscible materials systems, such as Au–Ag etc, have been studied in order to synthesize nanoparticles via PLiD.

Overall, the PLiD process has been recognized as an effective method to create nanoparticles of diverse elemental systems, as well as covering a large parameter space of nanoparticle sizes.

2.7 Hybrid dewetting

Laser dewetting has emerged as a versatile method for transforming thin metallic films into nanoparticles, offering a pathway to fabricate nanostructures with plasmonic, catalytic, and sensing functionalities. However, a persistent limitation of conventional laser dewetting is the broad particle size distribution typically obtained, which leads to multiple plasmon resonance modes and broad spectral linewidths. These features reduce the sensitivity and figure of merit of plasmonic devices, restricting their applicability in areas such as sensing and spectroscopy. The challenge, therefore, lies in developing strategies that preserve the simplicity and scalability of laser dewetting while simultaneously improving nanoparticle size uniformity.

A recent study demonstrated an effective solution through a hybrid dewetting approach that combines thermal annealing with pulsed laser irradiation, as shown in

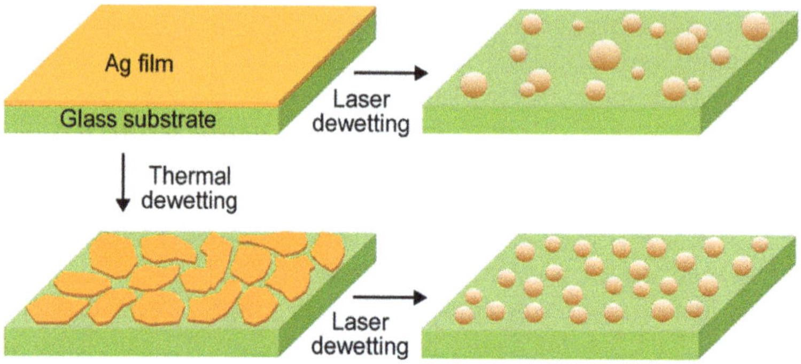

Figure 2.10. Hybrid dewetting process for generating uniform Ag nanoparticles. Reprinted from [51], Copyright (2021), with permission from Elsevier.

figure 2.10. In this method, a thin silver film of 10 nm thickness, deposited on a glass substrate, was first annealed at 250 °C in an inert argon atmosphere. This pre-annealing step induced solid-state dewetting, leading to an island-like morphology rather than a continuous film. When subsequently irradiated with nanosecond Nd: YAG laser pulses, these islands acted as predefined precursors for nanoparticle formation. As a result, the hybrid process produced Ag nanoparticles with significantly narrower size distribution compared to direct laser dewetting. Whereas laser irradiation alone yielded particles in the range of 35–145 nm, the hybrid approach generated nanoparticles with an average size of about 48 nm and a standard deviation of only 10.5 nm, as shown in figure 2.11.

The impact of improved size uniformity was most evident in the optical response of the nanoparticles. Direct laser dewetting resulted in two weak localized surface plasmon resonance peaks, corresponding to dipolar and quadrupolar modes associated with larger particles. In contrast, the hybrid-dewetted nanoparticles exhibited a strong and sharp single plasmon peak, reflecting the suppression of multipolar contributions due to the smaller and more uniform particle sizes, as shown in figure 2.11. This sharper resonance enhanced the refractive index sensitivity of the system, with the hybrid-dewetted particles achieving a sensitivity of 118 nm RIU^{-1} and a figure of merit of 2.145, values substantially higher than those typically reported for laser-dewetted Ag or bimetallic systems. Structural and chemical characterization confirmed that the particles were crystalline and unoxidized, highlighting the ability of the rapid heating and cooling cycle in laser dewetting to suppress oxidation.

Beyond optical sensing, the hybrid dewetting approach was also evaluated for surface-enhanced Raman scattering. Interestingly, the enhancement factor of the hybrid-dewetted nanoparticles was lower than that of direct laser-dewetted samples when probed at 532 nm, owing to the mismatch between the excitation wavelength and the sharper plasmon resonance of the hybrid particles. Nevertheless, this result underlines the tunability of hybrid dewetting: by adjusting film thickness or laser conditions to tailor resonance positions, the method can be adapted for optimized Raman enhancement. The study further confirmed that nanoparticle size scales with initial film thickness, providing an additional handle to tune optical properties.

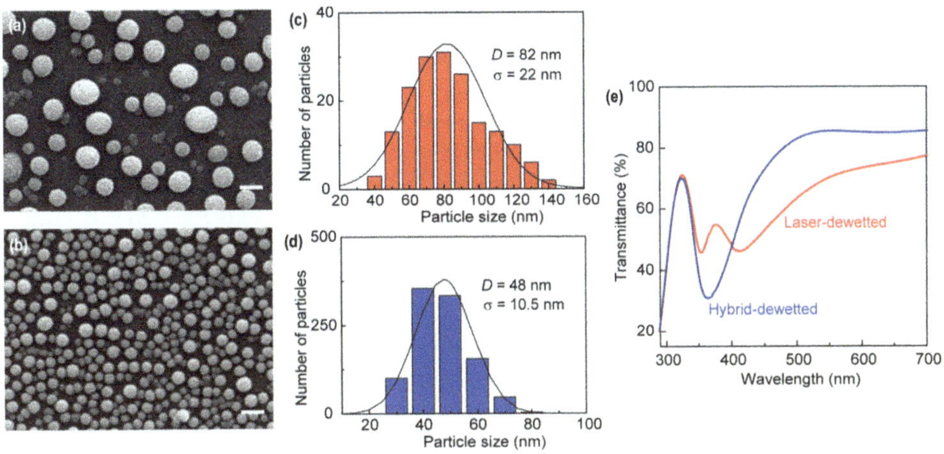

Figure 2.11. (a) Laser-dewetted particles. (b) Hybrid-dewetted particles. Scale bars are 100 nm. (c) and (d) Size distributions of the laser- and hybrid-dewetted particles, respectively. The solid line represents fitting to a normal distribution curve (D = mean size and σ = standard deviation). (e) Transmission spectra. Reprinted from [51], Copyright (2021), with permission from Elsevier.

Overall, the hybrid dewetting strategy demonstrates how a simple modification to conventional laser dewetting can overcome its intrinsic drawback of broad particle size distribution. By integrating a brief thermal annealing step prior to laser irradiation, it is possible to produce highly uniform plasmonic nanoparticles with superior optical performance, while retaining the lithography-free and scalable nature of laser processing. This development enriches the toolkit of laser-based nanostructure fabrication and opens opportunities for more reliable and sensitive plasmonic devices, particularly in refractive index sensing and potentially in spectroscopic applications once resonance conditions are matched.

Bibliography

[1] Kondic L, Gonzalez A G, Diez J A, Fowlkes J D and Rack P 2020 Liquid-state dewetting of pulsed-laser-heated nanoscale metal films and other geometries *Annu. Rev. Fluid Mech.* **52** 235–62

[2] Krishna H, Shirato N, Favazza C and Kalyanaraman R 2009 Energy driven self-organization in nanoscale metallic liquid films *Phys. Chem. Chem. Phys.* **11** 8136–43

[3] Vrij A and Overbeek J T G 1968 Rupture of thin liquid films due to spontaneous fluctuations in thickness *J. Am. Chem. Soc.* **90** 3074–8

[4] Reiter G 1992 Dewetting of thin polymer films *Phys. Rev. Lett.* **68** 75–8

[5] Thiele U, Velarde M G and Neuffer K 2001 Dewetting: film rupture by nucleation in the spinodal regime *Phys. Rev. Lett.* **87** 016104

[6] Ruffino F and Grimaldi M G 2019 Nanostructuration of thin metal films by pulsed laser irradiations: a review *Nanomaterials* **9** 1333

[7] Krishna H, Miller C, Longstreth-Spoor L, Nussinov Z, Gangopadhyay A K and Kalyanaraman R 2008 Unusual size-dependent magnetization in near hemispherical Co nanomagnets on SiO[sub 2] from fast pulsed laser processing *J. Appl. Phys.* **103** 073902

[8] Krishna H, Sachan R, Strader J, Favazza C, Khenner M and Kalyanaraman R 2010 Thickness-dependent spontaneous dewetting morphology of ultrathin Ag films *Nanotechnology* **21** 155601

[9] Yadavali S, Khenner M and Kalyanaraman R 2013 Pulsed laser dewetting of Au films: experiments and modeling of nanoscale behavior *J. Mater. Res.* **28** 1715–23

[10] Trice J, Thomas D, Favazza C, Sureshkumar R and Kalyanaraman R 2007 Pulsed-laser-induced dewetting in nanoscopic metal films: theory and experiments *Phys. Rev.* B **75** 235439

[11] van Benthem K 2015 Wetting and dewetting of ultra-thin Ni films on Si and SiO_2 substrates *Microsc. Microanal.* **21** 775–6

[12] Meshot E R, Zhao Z, Lu W and Hart A J 2014 Self-ordering of small-diameter metal nanoparticles by dewetting on hexagonal mesh templates *Nanoscale* **6** 10106–12

[13] Owusu-Ansah E, Birss V I and Shi Y 2020 Mechanisms of pulsed laser-induced dewetting of thin platinum films on tantalum substrates—a quantitative study *J. Phys. Chem.* C **124** 23387–93

[14] Bischof J, Scherer D, Herminghaus S and Leiderer P 1996 Dewetting modes of thin metallic films: nucleation of holes and spinodal dewetting *Phys. Rev. Lett.* **77** 1536–9

[15] Fowlkes J D, Kondic L, Diez J, Wu Y and Rack P D 2011 Self-assembly versus directed assembly of nanoparticles via pulsed laser induced dewetting of patterned metal films *Nano Lett.* **11** 2478–85

[16] Naffouti M *et al* 2017 Complex dewetting scenarios of ultrathin silicon films for large-scale nanoarchitectures *Sci. Adv.* **3** 1–11

[17] Yoo J H, In J B, Zheng C, Sakellari I, Raman R N, Matthews M J, Elhadj S and Grigoropoulos C P 2015 Directed dewetting of amorphous silicon film by a donut-shaped laser pulse *Nanotechnology* **26** 165303

[18] Sachan R, Ramos V, Malasi A, Yadavali S, Bartley B, Garcia H, Duscher G and Kalyanaraman R 2013 Oxidation-resistant silver nanostructures for ultrastable plasmonic applications *Adv. Mater.* **25** 2045–50

[19] Sachan R, Yadavali S, Shirato N, Krishna H, Ramos V, Duscher G, Pennycook S J, Gangopadhyay A K, Garcia H and Kalyanaraman R 2012 Self-organized bimetallic Ag–Co nanoparticles with tunable localized surface plasmons showing high environmental stability and sensitivity *Nanotechnology* **23** 275604

[20] Malasi A, Sachan R, Ramos V, Garcia H, Duscher G and Kalyanaraman R 2015 Localized surface plasmon sensing based investigation of nanoscale metal oxidation kinetics *Nanotechnology* **26** 205701

[21] Sachan R, Malasi A, Ge J, Yadavali S, Krishna H, Gangopadhyay A, Garcia H, Duscher G and Kalyanaraman R 2014 Ferroplasmons: intense localized surface plasmons in metal-ferromagnetic nanoparticles *ACS Nano* **8** 9790–8

[22] Wu Y, Fowlkes J D and Rack P D 2011 The optical properties of Cu–Ni nanoparticles produced via pulsed laser dewetting of ultrathin films: the effect of nanoparticle size and composition on the plasmon response *J. Mater. Res.* **26** 277

[23] Fowlkes J D, Wu Y and Rack P D 2010 Directed assembly of bimetallic nanoparticles by pulsed-laser-induced dewetting: a unique time and length scale regime *ACS Appl. Mater. Interfaces* **2** 2153–61

[24] Gangopadhyay A K, Krishna H, Favazza C, Miller C and Kalyanaraman R 2007 Heterogeneous nucleation of amorphous alloys on catalytic nanoparticles to produce 2D patterned nanocrystal arrays *Nanotechnology* **18** 485606

[25] Allaire R H, Kondic L, Cummings L J, Rack P D and Fuentes-cabrera M 2021 The role of phase separation on Rayleigh–Plateau type instabilities in alloys *J. Phys. Chem.* C **125** 5723–31
[26] Mckeown J T, Wu Y, Fowlkes J D, Rack P D and Campbell G H 2015 Simultaneous *in situ* synthesis and characterization of Co @ Cu core–shell nanoparticle arrays *Adv. Mater.* **27** 1060–5
[27] Krishna H, Shirato N, Yadavali S, Sachan R, Strader J and Kalyanaraman R 2011 Self-organization of nanoscale multilayer liquid metal films: experiment and theory *ACS Nano* **5** 470–6
[28] Yadavali S, Krishna H and Kalyanaraman R 2012 Morphology transitions in bilayer spinodal dewetting systems *Phys. Rev. B Condens. Matter Mater. Phys.* **85** 235446
[29] Krishna H, Gangopadhyay A K, Strader J and Kalyanaraman R 2011 Nanosecond laser-induced synthesis of nanoparticles with tailorable magneticanisotropy *J. Magn. Magn. Mater.* **323** 356–62
[30] Olbrich M, Punzel E, Lickschat P, Weißmantel S and Horn A 2016 Investigation on the ablation of thin metal films with femtosecond to picosecond-pulsed laser radiation *Phys. Proc.* **83** 93–103
[31] Kuznetsov A I, Koch J and Chichkov B N 2009 Nanostructuring of thin gold films by femtosecond lasers *Appl. Phys. A Mater. Sci. Process* **94** 221–30
[32] Mandal S, Gupta A K, Kone A, Hachtel J A and Sachan R 2024 Creation of multi-principal element alloy NiCoCr nanostructures via nanosecond laser-induced dewetting *Small* **20** 2309574
[33] Paduri V R, Harimkar S and Sachan R 2024 Bimetallic AgCo nanoparticle synthesis via combinatorial nanosecond laser-induced dewetting of thin films *Adv. Eng. Mater.* **26** 2401258
[34] Schaaf P 2010 *Laser Processing of Materials: Fundamentals, Applications and Developments* vol 139 (Heidelberg: Springer)
[35] Ivanov D and Zhigilei L 2003 Combined atomistic-continuum modeling of short-pulse laser melting and disintegration of metal films *Phys. Rev. B: Condens. Matter Mater. Phys.* **68** 064114
[36] Ruffino F and Grimaldi M G 2015 Controlled dewetting as fabrication and patterning strategy for metal nanostructures *Phys. Status Solidi* A **212** 1662–84
[37] Ruffino F and Grimaldi M G 2021 Nano-shaping of gold particles on silicon carbide substrate from solid-state to liquid-state dewetting *Surf. Interfaces* **24** 101041
[38] Veiko V P and Konov V I (ed) 2014 *Fundamentals of Laser-Assisted Micro-and Nanotechnologies* (Cham: Springer International Publishing)
[39] Allen P B 1987 Theory of thermal relaxation of electrons in metals *Phys. Rev. Lett.* **59** 1460
[40] Becker J, Grün G, Seemann R, Mantz H, Jacobs K, Mecke K R, Blossey R, Mathematik A, Bonn U and Bonn D 2003 Complex dewetting scenarios captured by thin-film models *Nat. Mater.* **2** 59–63
[41] Bonn D, Eggers J, Indekeu J and Meunier J 2009 Wetting and spreading *Rev. Mod. Phys.* **81** 739–805
[42] Israelachvili J 1997 Intermolecular & surface forces *J. Chem. Inf. Model.* **53** 1689–99
[43] Ajaev V S and Willis D A 2003 Thermocapillary flow and rupture in films of molten metal on a substrate *Phys. Fluids* **15** 3144–50

[44] Atena A and Khenner M 2009 Thermocapillary effects in driven dewetting and self assembly of pulsed-laser-irradiated metallic films *Phys. Rev. B: Condens. Matter Mater. Phys.* **80** 075402

[45] Mitlin V S 1994 On dewetting conditions *Colloids Surf. A: Physicochem. Eng. Asp.* **89** 97–101

[46] Luber E J, Olsen B C, Ophus C and Mitlin D 2010 Solid-state dewetting mechanisms of ultrathin Ni films revealed by combining *in situ* time resolved differential reflectometry monitoring and atomic force microscopy *Phys. Rev. B Condens. Matter Mater. Phys.* **82** 085407

[47] Favazza C, Kalyanaraman R and Sureshkumar R 2006 Robust nanopatterning by laser-induced dewetting of metal nanofilms *Nanotechnology* **17** 4229–34

[48] Shirato N, Krishna H and Kalyanaraman R 2010 Thermodynamic model for the dewetting instability in ultrathin films *J. Appl. Phys.* **108** 024313

[49] Krishna H, Shirato N, Favazza C and Kalyanaraman R 2011 Pulsed laser induced self-organization by dewetting of metallic films *J. Mater. Res.* **26** 154–69

[50] Krishna H, Favazza C, Gangopadhyay A K and Kalyanaraman R 2008 Functional nanostructures through nanosecond laser dewetting of thin metal films *JOM* **60** 37–42

[51] Oh H *et al* 2021 A hybrid dewetting approach to generate highly sensitive plasmonic silver nanoparticles with a narrow size distribution *Appl. Surf. Sci.* **542** 148613

Chapter 3

Laser-synthesis in liquid

This chapter reviews the principles, mechanisms, and advancements in laser ablation in liquids (LALs) for nanoparticle synthesis, chemical modification, and nanopatterning. LALs employ high-intensity pulsed lasers to ablate submerged solid targets, initiating complex laser–matter–liquid interactions that drive plasma formation, cavitation bubble dynamics, and nanoparticle nucleation. The liquid medium acts as both a confining and reactive environment, influencing particle size, morphology, and surface chemistry through solvent effects, reactive species generation, and bubble microreactor chemistry. Strategies for controlling nanoparticle properties include solvent selection, surfactant addition, and chemical additives to modulate oxidation, reduction, passivation, and phase control. Recent developments in hybrid LAL techniques—such as magnetic, electric, thermal, acoustic, photonic, and microfluidic assistance—offer enhanced scalability, morphology control, and compositional tuning. Beyond colloidal nanoparticle production, LAL enables high-resolution surface nanopatterning through liquid-mediated laser structuring, producing features such as periodic gratings, microgrooves, and nanocavities. This chapter highlights LAL's versatility in fabricating functional nanomaterials and patterned surfaces for applications in catalysis, photonics, biomedicine, and electronics.

3.1 Fundamental physics of laser ablation in liquids

Laser ablation in liquids is a versatile physical method for synthesizing nanomaterials, especially nanoparticles, by focusing high-energy laser pulses onto a solid target submerged in a liquid medium. The fundamental physics of LAL involves complex multi-scale interactions among laser light, the solid target, and the surrounding liquid, leading to material removal, plasma formation, and nanoparticle generation. This process differs significantly from ablation in vacuum or air due to the confining and reactive nature of the liquid environment. The liquid environment also allows the formation of nanoparticles in different surface stages, e.g., uncapped and capped nanoparticles with selected molecules (figure 3.1).

doi:10.1088/978-0-7503-6224-5ch3

Recently, increasing attention is being paid to reactive LAL, in which the LAL of 《A》 is performed with an additional additive 《B》 because it paves a new way for the synthesis of diverse multi-component 《A/B》 nanomaterials starting from pure bulk elemental targets and powered via their reactions with liquids (e.g., water, hydrogen peroxide, organic solvents, and superfluid/supercritical fluids) or liquid additives (e.g., salts, surfactant, support, colloid, and their mixtures).

3.1.1 Laser–matter interaction at the solid–liquid interface

When a high-intensity pulsed laser (typically nanosecond, picosecond, or femtosecond duration) irradiates a submerged target, the absorption of laser energy leads to rapid heating, melting, and vaporization of the surface layer. The rate and depth of energy deposition are governed by the laser wavelength, pulse duration, and target absorption characteristics [4].

In nanosecond LAL, thermal processes dominate, and material removal occurs via melting and explosive boiling. In contrast, ultrafast femtosecond LAL primarily involves non-thermal mechanisms like multiphoton ionization and Coulomb explosion, offering more controlled ablation with minimal thermal damage [2].

3.1.2 Plasma formation and shockwave generation

The LAL process begins with absorption of the laser pulse by the solid target, leading to rapid heating, melting, and vaporization of the surface layer. The intense energy density (10^8–10^{11} W cm^{-2}) initiates ionization of the ablated material, forming a dense plasma plume at the solid–liquid interface. This plasma contains atoms, ions, clusters, and electrons, expanding at supersonic velocities into the surrounding liquid. In the liquid environment, the plasma expansion is rapidly confined by the surrounding medium, generating strong pressure gradients that produce shockwaves and cavitation bubbles. The confinement also increases the plasma temperature and pressure, enhancing the energy density and leading to efficient material breakdown and nucleation of nanoparticles. These phenomena contribute to additional mechanical ablation of the target and influence the spatial distribution and size of the resulting nanoparticles.

3.1.3 Bubble dynamics and nanoparticle formation

The high-pressure plasma rapidly expands and forms a cavitation bubble, which undergoes oscillations and eventually collapses. During this time, the ejected atoms and clusters cool down via condensation and recombination, leading to the nucleation and growth of nanoparticles within the bubble or surrounding liquid, as shown in figure 3.1.

The cooling rate, nucleation kinetics, and interactions with reactive species in the liquid (e.g., surfactants, solvents, or ions) determine the size, shape, and composition of the synthesized nanoparticles. Repeated laser pulses can also fragment previously formed nanoparticles, enabling fine-tuning of their characteristics.

The bubble undergoes oscillatory dynamics consisting of expansion–collapse cycles. The expansion phase is driven by plasma pressure and vaporization of liquid

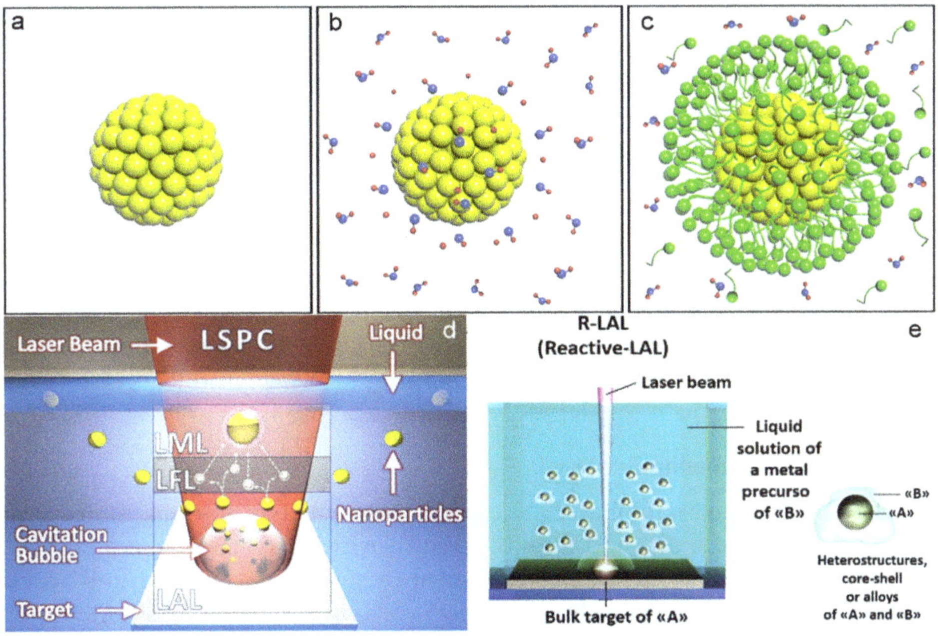

Figure 3.1. (a–c) Sketches of nanoparticles with various surface states: (a) naked nanoparticles in vacuum; (b) uncapped nanoparticles in water; (c) capped nanoparticles in an aqueous solution. Reprinted from [1], Copyright (2019), with permission from Elsevier. (d) Scheme of laser melting/fragmentation in liquids (LML/LFL) involving laser ablation in liquids (LAL). LSPC: laser synthesis and processing of colloid in liquids. Reprinted with permission from [2]. Copyright (2017) American Chemical Society. (e) Sketch of reactive LAL (R-LAL) using a target made of material 《A》 to give nanoparticles that interact with a chemically reactive precursor of 《B》 to give a multi-component 《A/B》 system. Reproduced from [3]. CC BY 4.0.

at the interface. This stage disperses ablated species into the bubble volume. Whereas, in the collapse phase, the bubble implodes as the pressure equalizes with the liquid phase, producing secondary shockwaves and sometimes splitting into daughter bubbles. These collapses can occur multiple times (typically 2–4 cycles), each redistributing nanoparticles and clusters. Inside the bubble, temperature gradients (thousands of Kelvin at the core versus ambient at the periphery) and high-pressure conditions ($\sim$100 bar) govern nucleation. The bubble wall oscillation also entrains surrounding liquid, promoting quenching and nanoparticle passivation.

The key factors that prominently control the bubble dynamics and nanoparticle formation are as follows:

- **Laser parameters:** Pulse energy, fluence, and duration determine plasma density and thus initial bubble size. Femtosecond pulses generally produce smaller, more monodisperse nanoparticles compared to nanosecond pulses due to reduced heat diffusion.
- **Liquid environment:** Viscosity, surface tension, and dissolved gas content influence bubble lifetime and stability. Polar solvents (e.g., water, alcohols) yield different nanoparticle sizes than nonpolar solvents.

- **Repetition rate:** High repetition rates can lead to bubble shielding, where residual bubbles from earlier pulses absorb or scatter laser energy, modifying ablation efficiency.
- **Target composition:** Thermal conductivity and optical absorption of the target dictate plasma expansion dynamics and subsequent bubble characteristics.

Among these, laser parameters and repetition rate determine the structure and morphology, while target composition governs the composition and combinations of constituent elements in the nanoparticles. In contrast, liquid media in LAL plays an important role in nanoparticle size, as detailed in the following section, by controlling the bubble lifetime, surface state, and overall composition, as it can enable various reactions leading to the formation of oxides, nitrides, etc.

3.1.4 Role of liquid medium

The laser synthesis of nanomaterials can be performed in various atmospheres, including vacuum, gas, air, and liquids, each of which has its pros and cons. The materials synthesized in ultrahigh vacuum are extraordinarily pure and reactive because of their naked surfaces, but expensive vacuum equipment is required, which will be a heavy economic burden to researchers who are interested in this process. Air is the cheapest atmosphere for laser ablation, but nanomaterials suspended in the air may be very toxic for human beings, as well as having limited control on morphology. Liquids have been proven to be the most favorable environment for nanomaterial synthesis in terms of spontaneous material dispersion, the inhibition of nanomaterial air pollution, and the availability of a large variety of uncapped and capped products and facilitated by the chemical interactions between target materials and liquids.

The liquid medium in LAL plays a dual role as both a confining environment and an active chemical participant. When the laser pulse ablates the solid target, the surrounding liquid confines the expanding plasma plume, producing steep pressure and temperature gradients that initiate cavitation bubble formation. This confinement leads to rapid quenching rates (10^9–10^{11} K s^{-1}), favoring the nucleation of nanometer-sized clusters and preventing excessive coarsening. At the same time, the viscosity, density, and surface tension of the liquid dictate the lifetime and oscillation dynamics of the cavitation bubble, which in turn control particle size distribution, morphology, and crystallinity. For example, high-viscosity solvents slow bubble collapse and promote larger nanoparticles, whereas low-viscosity, volatile solvents favor finer dispersions.

Beyond physical confinement, the liquid medium also governs the surface chemistry and final composition of nanoparticles. Reactive solvents such as water or alcohols provide radicals (H, OH·) that oxidize or hydroxylate nanoparticle surfaces, stabilizing them against aggregation. Alternatively, ionic liquids, surfactant-containing media, or organic solvents can cap, passivate, or even alloy the ablated species, yielding tailored surface functionality. Thus, by carefully choosing and engineering the liquid environment, researchers can tune nanoparticle size, dispersity, phase composition, and surface chemistry, making the liquid medium a critical design parameter for controlled nanoparticle synthesis in LAL. For example, polar solvents (e.g., water, ethanol) and nonpolar solvents (e.g., toluene) influence not only heat transfer and bubble dynamics but also surface passivation and oxidation states of the nanoparticles. The presence of surfactants or dissolved gases can further modify the nanoparticle size and prevent agglomeration [5].

An important example of the role of liquid media is presented by Liu *et al,* as shown in figure 3.2. This figure illustrates the time-evolution of Te nanostructures synthesized by LAL and highlights the critical role of the liquid medium in directing nanoparticle assembly pathways. On the left, a schematic summarizes the kinetic sequence: immediately after ablation, uncapped Te nanoparticles form due to rapid quenching of the ablation plume. Over time, these nanoparticles undergo crystallization and coalescence, followed by slow aggregation or self-assembly into one-dimensional Te nanochains. With further aging, the system transitions into larger Te agglomerates and, under extremely slow recrystallization, ultimately produces spherical microparticles. This cascade emphasizes how LAL begins with ultrafast nucleation but evolves through slower processes controlled by the surrounding liquid environment.

The right-hand panel shows TEM/SEM images comparing nanoparticle evolution in different solvents: H_2O, CH_3OH, CH_3CH_2OH, CH_3COCH_3, and CH_2Cl_2. In all solvents, small nanoparticles appear within the first seconds of ablation (row 1). However, their subsequent transformations differ markedly depending on solvent polarity, viscosity, and reactivity. For example, in water, nanoparticles quickly assemble into chains (A2, 10 s) and then irregular agglomerates (A3, 1 h),

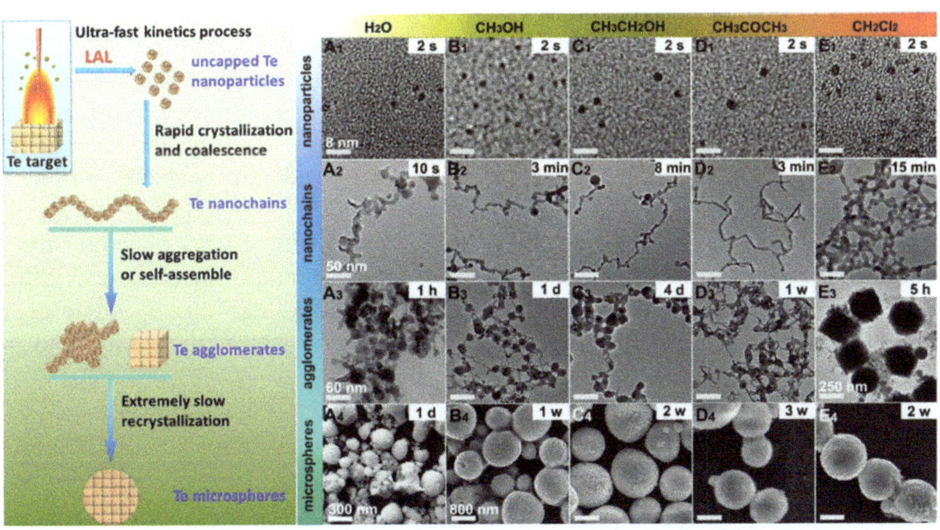

Figure 3.2. Schematic and TEM images of the spontaneous growth of uncapped Te nanoparticles in H_2O, CH_3OH, CH_3CH_2OH, CH_3COCH_3, and CH_2Cl_2 at different aging periods. Scale bar: 8 nm for (A1–E1), 50 nm for (A2–E2), 60 nm for (A3–D3), 250 nm for (E3), 300 nm for (A4), and 800 nm for (B4–E4). The aging time is marked in each figure. Reproduced from [6]. CC BY 4.0.

ultimately recrystallizing into microspheres within a day (A4). In contrast, methanol and ethanol slow down this progression, with chain and agglomerate stages persisting for minutes to days before microsphere formation (B4–C4). Aprotic solvents such as acetone and dichloromethane further delay or alter these transitions: acetone allows extended chain persistence (D2–D3) before eventual microsphere formation after weeks, while dichloromethane yields unusually large agglomerates (E3, 5 h) and delayed microspheres (E4, 2 weeks).

Together, these results demonstrate how the liquid medium controls nanoparticle stability, aggregation kinetics, and recrystallization pathways. Polar solvents with strong hydrogen bonding (e.g., water, alcohols) promote rapid chain and sphere formation, whereas nonpolar or less polar solvents slow down assembly and stabilize intermediate structures. Thus, by tuning solvent properties, one can steer Te nanostructures toward desired morphologies—ranging from transient nanoparticles and chains to stable microspheres. This highlights the liquid medium not only as a passive confinement environment in LAL but as an active director of nanomaterial evolution through solvation, surface passivation, and aggregation dynamics.

3.1.5 Timescales and energy transfer

LAL is a highly dynamic process governed by ultrafast energy deposition and subsequent multiscale phenomena. The sequence begins on the femtosecond to picosecond timescale, where the laser pulse is absorbed by the solid target, causing rapid electronic excitation and non-equilibrium heating of the surface layer. In this regime, electron–phonon coupling transfers energy from hot carriers to the lattice, setting the stage for melting and vaporization. The progression as a function of timescale is illustrated in figure 3.3.

On the picosecond to nanosecond scale, this localized energy deposition drives phase transitions, material ejection, and the formation of a dense plasma plume at the solid–liquid interface. The plume expands violently, transferring momentum to the surrounding liquid and generating shockwaves. Concurrently, the high cooling rates initiate primary nucleation of atomic clusters and nanoparticles within the confined plasma.

Over nanoseconds to microseconds, the expansion of the plasma initiates cavitation bubble formation. These bubbles entrap ablated species, providing a unique microreactor environment where nanoparticles can grow, coalesce, or aggregate. Bubble oscillations further influence nanoparticle dispersity and size

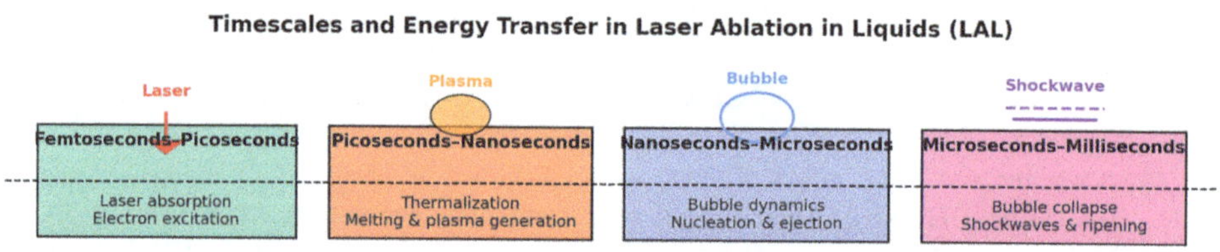

Figure 3.3. An illustration of the timescale and energy transfer at various stages of LAL.

distribution, while plume–bubble interactions determine how much material is redeposited or expelled.

Finally, on the microsecond to millisecond timescale, cavitation bubble collapse and repeated oscillations release shockwaves back into the liquid. These processes redistribute nanoparticles, promote secondary nucleation, and govern diffusion-mediated growth and ripening. The overlapping timescales highlight how ultrafast laser–matter interactions are coupled to slower hydrodynamic and chemical events, collectively determining the morphology, crystallinity, and surface chemistry of LAL-synthesized nanomaterials.

3.2 Chemical interactions

LAL technique provides a clean and versatile approach to synthesize a wide range of nanoparticles without the need for chemical precursors or stabilizing agents. However, despite being fundamentally a physical process, chemical interactions between the ablated species and the surrounding liquid environment play a critical role in determining the composition, size, surface chemistry, and long-term stability of the synthesized nanoparticles [7–9].

3.2.1 Solvent decomposition and reactive species generation

Upon laser irradiation, the high energy density at the target–liquid interface not only causes material ejection but also induces localized decomposition of the liquid medium. This can generate a variety of reactive species, such as radicals, ions, or solvated electrons. For instance:

- In water, photolysis and pyrolysis may produce hydroxyl radicals (·OH), hydrogen atoms (H·), and solvated electrons (e^{-}_aq) [10, 11].
- In organic solvents like ethanol or acetone, laser-induced photochemical reactions can generate carbonaceous fragments and reducing species [12].

These species can react with the freshly ablated metal atoms or clusters, promoting oxidation, reduction, or surface passivation, depending on the local environment. Zhang *et al* demonstrated a promising strategy for fabricating Fe–N–C/rGO electrocatalysts through the *ex situ* integration of LAL-synthesized FeO colloids with GO nanosheets, followed by polymerization with pyrrole monomers and subsequent pyrolysis [13]. The FeO colloids were prepared via LAL of iron in acetone, yielding stable nanoparticle suspensions. Upon integration and carbonization, the resulting Fe–N–C/rGO hybrid exhibited oxygen reduction reaction activity comparable to that of commercial Pt/C catalysts, attributed to synergistic factors such as N-doping-induced electronic modulation, efficient charge transfer between the catalyst supports and reactants, a high surface area, and an abundance of accessible active sites [13].

Building upon this strategy, the synthesis of Co@NC/rGO composite catalysts, comprising Co-based nanoparticles encapsulated in nitrogen-doped graphitic carbon layers supported on reduced GO, was achieved via *ex situ* mixing of LAL-generated Co colloids with GO sheets, followed by hydrothermal treatment and

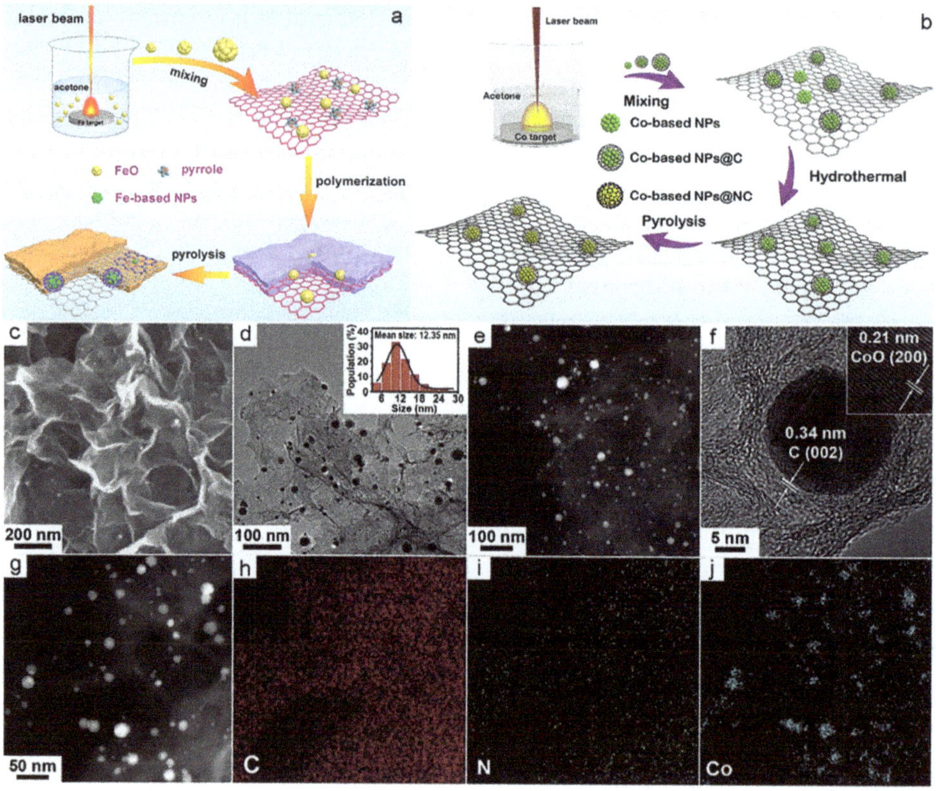

Figure 3.4. Polymerization and hydrothermal post-treatments of *ex situ* LAL-synthesized mixed colloid solutions for doped and complex supported-composite synthesis. (a) Schematic of *ex situ* LAL followed by polymerization and pyrolysis for Fe–N–C/rGO catalyst synthesis. Reprinted with permission from [13]. Copyright (2018) American Chemical Society. (b) Schematic of *ex situ* mixing followed by hydrothermal treatment and pyrolysis for the synthesis of rGO-supported Co-based nanoparticles encapsulated in nitrogen-doped carbon shells. (c) SEM, (d) TEM, (e) high-angle annular dark-field scanning transmission electron microscopy (HAADF-STEM) and (f) HRTEM images of Co@NC/rGO composites. (g) HAADF-STEM image and (h–j) EDS elemental C, N, and Co mappings of Co@NC/rGO composites. Reprinted from [14], Copyright (2020), with permission from Elsevier.

pyrolysis in a nitrogen atmosphere [14]. As characterized by SEM and TEM (figure 3.4), the Co nanoparticles were uniformly distributed across the rGO surface, exhibiting an average size of ∼12.35 nm. High-resolution lattice fringes with interplanar spacings of 0.21 and 0.34 nm corresponded to the (200) and (002) planes of CoO/C, confirming the formation of carbon-encapsulated CoO structures. EDS mapping further verified the homogeneous distribution of nitrogen and cobalt across the carbon matrix.

3.2.2 Surface passivation and stabilization

The interactions between reactive species and ablated particles influence surface chemistry and prevent uncontrolled agglomeration. The nature of the solvent—its polarity, dielectric constant, viscosity, and redox potential—thus directly governs the type and extent of surface reactions, which in turn impact nanoparticle dispersibility and reactivity [7, 15]. There are several studies taking advantage of the chemical interactions of liquid media to change the chemistry of nanoparticles. For example, metal nanoparticles tend to oxidize and form stable oxide shells (e.g., TiO_2, Fe_3O_4), affecting their functionality and optical properties [16, 17]. In another example, in ethanol, carbonaceous byproducts from solvent degradation can adsorb on the nanoparticle surface, forming protective organic layers that enhance colloidal stability [18]. Figure 3.5 demonstrates the influence of the liquid environment on bubble dynamics and nanoparticle formation during laser LAL, using a bulk AuFe alloy target ablated in different ethanol-based solutions.

In figure 3.5(a), the schematic highlights the ablation and nucleation process. A femtosecond or nanosecond laser pulse generates a plasma plume at the alloy surface, ejecting AuFe nuclei into the surrounding medium. In ethanol/H_2O_2, the strong oxidizing environment produces a larger cavitation bubble, allowing AuFe nanoparticles nanoparticles to nucleate and grow extensively within the confined space. This results in relatively larger and more polydisperse particles. In contrast, in ethanol/H_2O, the moderately oxidizing environment restricts bubble expansion and leads to the formation of smaller nanoparticles decorated with oxide crescents, arising from partial oxidation during bubble collapse. Figures 3.5(b) and (c) provide TEM evidence of these differences: ethanol/H_2O_2 leads to bigger, more spherical nanoparticles (tens of nanometers up to ~100 nm), while ethanol/H_2O yields smaller, less uniform particles with oxidized features. The size distribution histograms in panels d and e quantitatively confirm these trends. In ethanol/H_2O_2, the

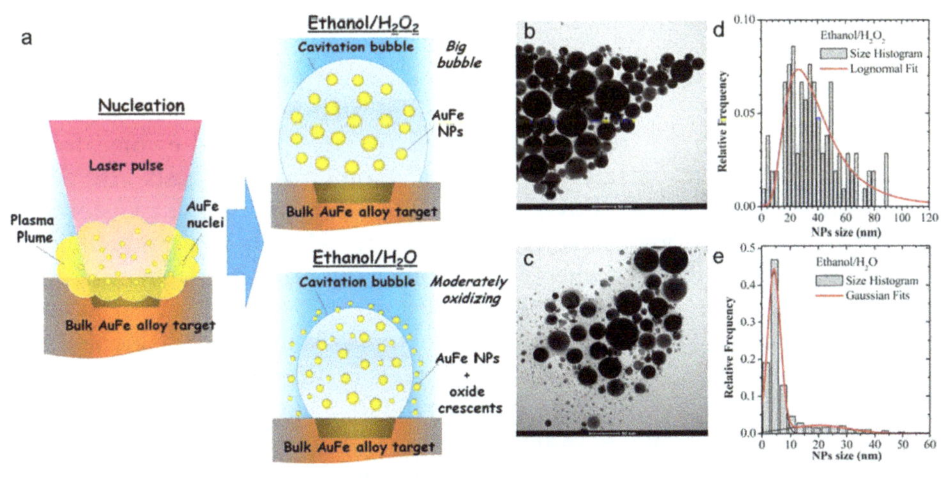

Figure 3.5. Liquid mixtures used for the LAL synthesis of AuFe nanoalloys. (a) Schematic of the formation of different sizes of AuFe nanoalloy by the LAL of AuFe in ethanol–H_2O_2 and ethanol–H_2O mixtures. (b) and (c) TEM images of AuFe nanoparticles synthesized in ethanol–H_2O_2 and ethanol–H_2O mixtures, respectively. (d) and (e) size distribution histograms of AuFe nanoparticles synthesized in ethanol–H_2O_2 and ethanol–H_2O mixtures, respectively. Reproduced from [40] with permission from the Royal Society of Chemistry.

distribution fits a lognormal curve, showing a broad range with a peak around 30–40 nm and a long tail extending to ~100 nm. In ethanol/H_2O, however, the histogram fits a narrow Gaussian distribution, peaking at smaller sizes (< 10 nm) with minimal broadening.

Overall, the figure demonstrates that the chemical nature of the liquid strongly influences cavitation bubble size, redox conditions, and nanoparticle growth pathways. A strongly oxidizing medium (H_2O_2) promotes rapid growth and coalescence of larger AuFe nanoparticles, while a moderately oxidizing medium (H_2O) constrains growth and favors smaller particles with surface oxidation. This underscores the dual role of the liquid in LAL: providing both physical confinement through bubble dynamics and chemical control through solvent reactivity.

3.2.3 Redox reactions and phase control

Chemical interactions can also drive redox transformations of the ablated species. In cases where the target is a multivalent metal (e.g., Fe, Mn, Ce), the oxidation state of the resulting nanoparticles can be altered by the oxidizing or reducing power of the medium. By adjusting the solvent composition or adding specific redox agents, it is possible to control the valence state, phase purity, or core–shell structure of the nanoparticles [19, 20].

For example:

- Au or Ag targets ablated in H_2O_2-containing solutions lead to controlled oxidation and formation of metal-oxide hybrid nanostructures [21].
- The addition of ligands or chelating agents (e.g., citrates, thiols) can selectively bind to nanoparticle surfaces, influencing nucleation pathways and shape evolution [22].

In another example, by altering the ratio of water and ethanol media, phase modulation of FeMn and MnO_2 nanoparticles has been reported by Zhang *et al* [2]. The study suggests that a reaction between Mn atoms and Fe ions, followed by surface oxidation, results in nonstoichiometric synthesis of Fe-rich FeMn@$FeMn_2O_4$ core–shell alloy particles. Interestingly, a phase transformation from Mn_3O_4 to Mn_2O_3 and finally to Ramsdellite γ-MnO_2 is accompanied by a morphology change from nanosheets to nanofibers in gradually increasing oxidizing environments.

Figure 3.6 illustrates the structural identification of manganese oxide nanofibers synthesized by laser ablation of FeMn alloy targets in different liquid environments. The selected area electron diffraction patterns (figure 3.6(a–c)) clearly show that the nanofibers obtained in ethanol–water mixtures (2:1 and 1:2) and in pure water correspond to different MnOx phases. Specifically, the nanofibers produced in ethanol–water (2:1) were indexed to Mn_3O_4, those from ethanol–water (1:2) to Mn_2O_3, and those in pure water to Ramsdellite γ-MnO_2. This systematic evolution indicates a phase transformation sequence $Mn_3O_4 \rightarrow Mn_2O_3 \rightarrow \gamma$-$MnO_2$ as the oxidizing power of the solvent increases with higher water content.

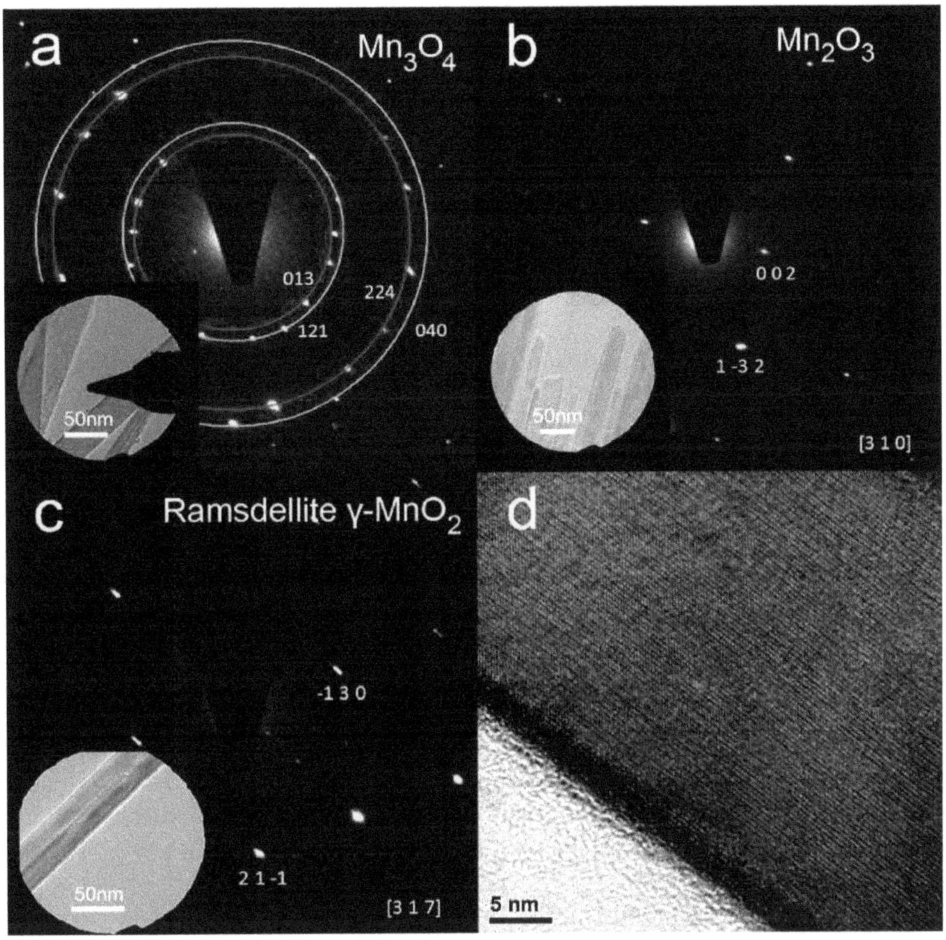

Figure 3.6. (a–c) Selected area electron diffraction patterns of the nanofibers obtained by LAL from a FeMn alloy target in ethanol–water mixtures with volume ratios of (2:1) and (1:2) and in pure water, respectively. (d) HRTEM image of a nanofiber showing the single-crystalline structure. [41] John Wiley & Sons. © 2017 WILEY-VCH Verlag GmbH & Co. KGaA, Weinheim

The high-resolution TEM image (figure 3.4(d)) further confirms that the nanofibers are highly crystalline and single-phase. The slight deviations in lattice parameters compared to ICSD database values were attributed to minor ($\sim$1%) Fe doping, which occurs due to the presence of Fe species in the ablation plume. Importantly, these results demonstrate that LAL can drive both morphology changes (from nanosheets to nanofibers) and phase transformations in manganese oxides simply by tuning the liquid medium. Thus, figure 3.6 underscores the strong influence of solvent composition on the oxidation environment during ablation, enabling precise control over the crystal phase and structure of MnOx nanomaterials.

3.2.4 Bubble chemistry and confinement effects

The cavitation bubble formed during LAL acts as a transient microreactor where chemical reactions can be significantly different from bulk liquid conditions. The high temperature (up to several thousand *K*) and pressure (hundreds of bars) inside the bubble enable unique reaction pathways that govern nanoparticle nucleation and growth [17]. Within this confined space:

- Metal atoms and ions can interact with solvent fragments.
- Nanoparticles may form through condensation, coalescence, or chemical reduction mechanisms [23].

The chemical microenvironment within the bubble thus strongly modulates nanoparticle composition and crystallinity.

3.2.5 Additives and surfactants

Deliberate addition of surfactants, stabilizers, or pH modifiers further enhances control over nanoparticle formation. Surfactants adsorb onto particle surfaces, modifying surface energy and growth directionality. Adjusting pH can influence the zeta potential and aggregation behavior [2, 24]. These secondary chemical interactions are crucial for tailoring nanoparticle size distribution, morphology, and long-term stability in colloidal suspensions.

Chemical interactions in LAL are not merely secondary phenomena but actively govern the nucleation, surface chemistry, and stability of the resulting nanoparticles. By tuning solvent composition, additives, and reaction conditions, one can harness these interactions to engineer functional nanomaterials with tailored properties for diverse applications in catalysis, biomedicine, and electronics.

3.3 Modifications in laser ablation in liquid methods

While LAL is a powerful standalone technique for nanoparticle synthesis, its scalability, precision, and control over particle properties can be significantly enhanced through externally-assisted hybrid approaches. These modifications integrate secondary energy sources, fields, or chemical environments into the LAL process, offering additional degrees of control over plasma dynamics, nucleation pathways, and material characteristics (figure 3.7).

3.3.1 Magnetically-assisted LA

The application of static or pulsed magnetic fields during LAL alters plasma confinement, bubble dynamics, and particle transport. Magnetic fields influence the motion of charged species in the laser-induced plasma, which can:

- increase plasma lifetime and temperature,
- enhance confinement and reduce nanoparticle agglomeration,
- promote anisotropic particle growth or alignment (e.g., nanorods, chains) [25].

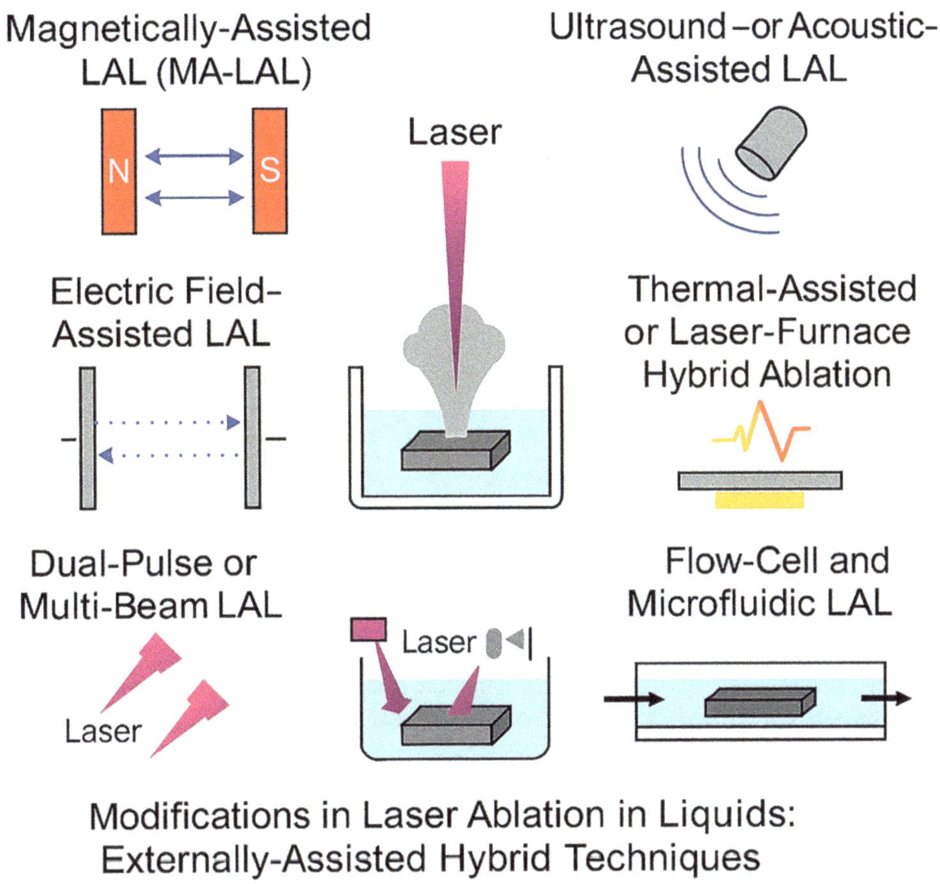

Figure 3.7. Illustration of externally-assisted hybrid techniques in LAL. The central schematic shows a conventional LAL setup where a pulsed laser ablates a submerged solid target. Surrounding modifications include: (i) magnetically-assisted LAL, where magnetic fields confine plasma and influence particle morphology; (ii) electric field-assisted LAL, which steers charged species and promotes anisotropic growth; (iii) dual-pulse or multi-beam LAL for enhanced ablation control and nanoparticle tailoring; (iv) ultrasound- or acoustic-assisted LAL, enhancing cavitation dynamics and dispersion; (v) thermal-assisted or laser–furnace hybrid ablation, enabling higher crystallinity and phase control; and (vi) flow-cell or microfluidic LAL, offering continuous and scalable nanoparticle synthesis.

This method is especially useful for tailoring magnetic nanoparticle morphology and magnetic properties, such as coercivity and saturation magnetization.

3.3.2 Ultrasound- or acoustic-assisted LAL

Acoustic cavitation, when combined with LAL, enhances the dispersion of nanoparticles and reduces agglomeration by:

- improving fluid mixing and bubble collapse symmetry,
- promoting secondary nucleation events,
- assisting in nanoparticle fragmentation and size reduction [26].

Ultrasound-assisted LAL is a promising route for size-controlled colloids, especially in the synthesis of noble metal and oxide nanoparticles.

3.3.3 Electric field-assisted LAL

Applying an external electric field across the ablation chamber modulates the migration of charged plasma species and influences:

- nanoparticle size distribution through electrostatic control,
- directional ejection or orientation of anisotropic structures,
- enhanced formation of core–shell or doped structures via field-driven ion accumulation [27].

This approach is useful in the *in situ* doping of semiconductors or the controlled growth of heterostructures.

3.3.4 Thermal-assisted or laser–furnace hybrid ablation

Combining LAL with a pre-heated target or solution alters the thermal environment of the ablation site:

- enhances melting and diffusion processes,
- promotes larger nanoparticle formation due to slower quenching,
- enables controlled crystallization during bubble collapse [28].

Such hybrid setups are useful for producing single-crystalline or phase-pure nanomaterials where high crystallinity is essential.

3.3.5 Dual-pulse or multi-beam LAL

Using two laser pulses (collinear or orthogonal, with femto–nano or pico–nano combinations) enhances ablation efficiency and allows temporal control of energy delivery:

- The first pulse initiates ablation; the second controls plasma dynamics or bubble collapse,
- enabling tunable fragmentation or reshaping of particles in real-time [3].

This method is particularly effective for surface structuring or tuning nanoparticle size distribution.

3.3.6 Photonic and plasmonic hybrid LAL

By coupling LAL with plasmonic substrates or focusing near a nanostructured interface, localized field enhancements can:

- lower ablation thresholds,
- promote localized heating or nonlinear optical effects,
- facilitate selective ablation and localized growth [29].

This technique is beneficial for site-specific synthesis or heterostructured nanoparticle formation.

3.3.7 Flow-cell and microfluidic LAL

Integrating LAL into microfluidic reactors or flow cells offers enhanced control over:

- residence time of nanoparticles,
- local chemical environment and mixing conditions,
- scalability and reproducibility for industrial applications [30, 31].

Microfluidic LAL is a promising platform for continuous and scalable nanoparticle production, especially for biomedical and catalytic applications.

Externally-assisted hybrid LAL techniques expand the functional versatility of conventional LAL, enabling greater control over nanoparticle composition, size, structure, and functionality. By synergistically integrating magnetic, electric, thermal, acoustic, or photonic fields, these methods address existing limitations of LAL and pave the way for more advanced, tunable nanomaterial synthesis.

3.4 Nanopatterning in liquid

LAL is traditionally recognized for nanoparticle synthesis; however, it also serves as a versatile method for nanopatterning and surface texturing, leveraging the confined ablation environment and laser–matter–liquid interactions to create periodic and aperiodic surface structures with tunable features.

3.4.1 Mechanism of nanopatterning in LAL

When a pulsed laser is focused onto a solid target immersed in a liquid, the intense localized energy induces melting, vaporization, and plasma formation. In addition to nanoparticle ejection, under specific fluence and pulse regimes, self-organized patterns such as:

- laser-induced periodic surface structures (LIPSS),
- microgrooves, and
- nanocavities

can emerge on the target surface [32, 33]. When a high-intensity pulsed laser irradiates a solid target in air, vacuum, or liquid, the interaction of the incident light with surface electromagnetic waves can lead to the spontaneous formation of LIPSS, often referred to as 'ripples.' These nanoscale patterns emerge due to interference between the incoming laser beam and surface-scattered waves, resulting in a periodic modulation of absorbed energy and localized melting. A simple schematic of the LIPSS process is shown in figure 3.8 [34].

The characteristic period of LIPSS is typically close to or slightly smaller than the laser wavelength, and its orientation depends on the laser polarization. Under femtosecond pulses, ultrafast electron–phonon coupling and non-equilibrium

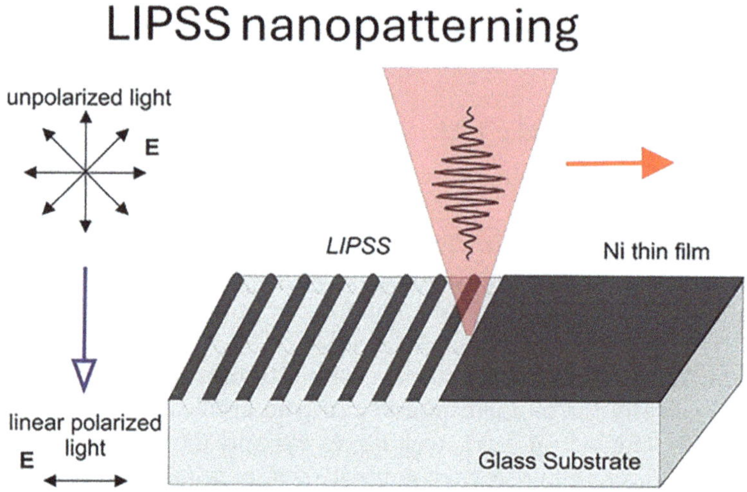

Figure 3.8. Schematic of the LIPSS process as shown by E Skoulas *et al.* Reprinted from [34], Copyright (2021), with permission from Elsevier.

dynamics further enhance pattern fidelity, enabling reproducible nanostructures without lithography.

In liquids, LIPSS formation is strongly coupled with bubble dynamics and nanoparticle redeposition. The confining effect of the liquid alters plume expansion and heat dissipation, which can either suppress or enhance ripple formation. Additionally, nanoparticles generated by LAL can redeposit onto the laser-structured surface, leading to hybrid nanopatterns composed of ripples decorated with nanoparticles or nanoclusters. Such liquid-assisted LIPSS not only provide increased surface area but also introduce hierarchical roughness and compositional gradients. These properties are valuable for tailoring wettability, adhesion, tribology, catalysis, and sensing. Thus, LIPSS nanopatterns represent a self-organized route to functional nanostructures, and when combined with LAL-driven nanoparticle formation, they offer a versatile pathway toward designing hierarchical, multifunctional surfaces.

Figure 3.9 shows SEM images of LIPSS formed on 100 nm Ni thin films deposited on different dielectric substrates (SiO_2 and sapphire). The panels demonstrate that under optimized irradiation conditions ($\lambda = 513$ nm, repetition rate 60 kHz, fluence 0.17 J cm^{-12}, scan speed 30 mm s^{-1}, and line spacing $\delta = 6$ μm), highly ordered and uniform nanowire-like ripples are produced across centimeter-scale areas. On both SiO_2 and Al_2O_3 substrates, the resulting arrays exhibit consistent orientation dictated by the laser polarization, with periodicities of about 475 ± 41 nm and average widths of 182 ± 66 nm. The figure highlights the scalability and reproducibility of the approach, where scanning of overlapping laser lines enables homogeneous nanopatterning over extended surfaces. The optical photograph in figure 3.9 further emphasizes that such nanostructured regions can cover areas as large as 1 cm^2 without noticeable loss of uniformity. These results confirm

Figure 3.9. Top-view SEM image area scans with LIPSS on Ni films for SiO_2 (a, b) and Al_2O_3 (c, d) fabricated with $\lambda = 513$ nm, $R = 60$ kHz, $\varphi = 0.17$ J cm^{-2}, $v = 30$ mm s^{-1}, and $\delta = 6$ µm. A physical photograph of (c, d) 1 cm^2 on Al_2O_3 substrate is presented in (e). Reprinted from [34], Copyright (2021), with permission from Elsevier.

that femtosecond-laser-induced self-organization provides a direct, single-step pathway to fabricate large-area wire-grid-like nanopatterns, essential for developing practical polarizing plates and metasurface-based optical elements.

In the LAL environment, the liquid confinement introduces unique features to pattern formation:

- Bubble dynamics and recoil pressure influence surface reorganization.
- Shockwave reflections from the liquid–plasma interface modulate energy redistribution.
- Plasma–liquid interactions can locally etch or redeposit species on the target surface.

These processes combine to create high-resolution patterns with periodicities ranging from tens of nanometers to microns, depending on laser parameters (wavelength, fluence, pulse duration), liquid properties, and scan strategies.

In an interesting example, Duocastella *et al* utilized the liquid droplet as a microlens to produce sub-wavelength nanoscale patterns on Si after laser irradiation [35]. Figure 3.10 illustrates a two-step droplet-assisted laser processing (DALP) that addresses the challenges encountered in current micro and nanopatterning techniques. In the presented two-step approach, first, laser-induced forward transfer is used to print droplets with sizes ranging from 10 to 200 µm into specific areas of interest of a surface. Next, the printed droplets are used as lenses to focus laser pulses directly below each droplet, locally modifying the surface. Notably, the same laser source used for printing can be used for surface patterning.

One can consider the droplet to act as the liquid version of solid-immersed lenses without the constraints involved in their fabrication and placement—droplets present a flawless surface and can be easily printed at desired locations on a surface.

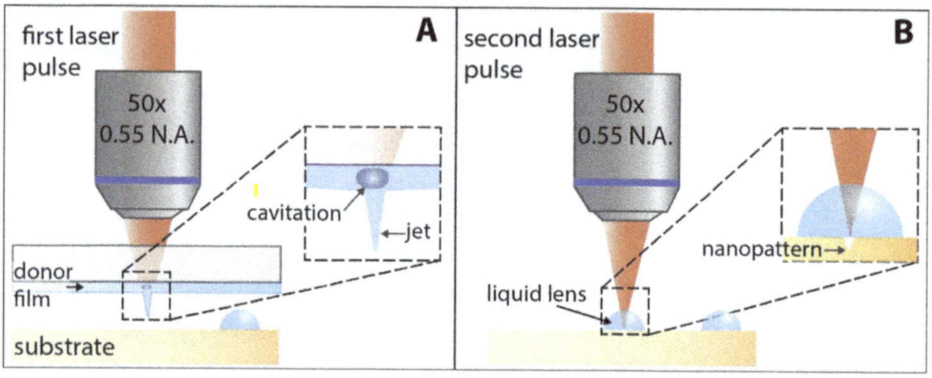

Figure 3.10. Schematic of the droplet-assisted laser processing approach. (A) A liquid droplet is printed on top of the surface to be patterned by using laser-induced forward transfer. (B) After removal of the donor film, the same laser source is focused through the droplet, which acts as a liquid lens. As a consequence, nanopatterns can be fabricated. Reproduced from [35]. CC BY 4.0.

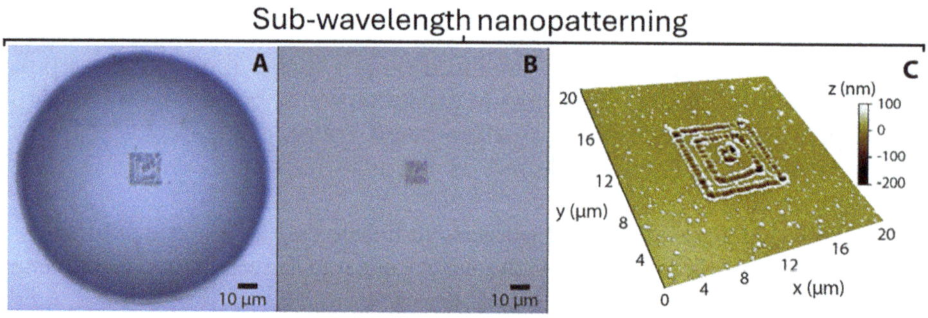

Figure 3.11. Nanopatterning of concentric squares. Optical micrograph of a pattern fabricated with a 100 μm droplet, (A) with the droplet still in place and (B), without the droplet. The drop produces a magnification of about a factor of 1.5. (C) Atomic force microscopy characterization of the pattern [35].

Thereby, this approach produces a straightforward increase in the focusing capabilities of a system by a factor that depends to first order on the liquid refractive index. In addition, contrary to immersion objectives, liquid micro-lenses can be used in a variety of optical systems, the focusing enhancement can be controlled by simply changing the liquid, and the small size of the lens minimizes possible absorption effects and even temporal dispersion in the case of irradiation by ultrafast laser pulses. This approach demonstrates the feasibility of creating nano-patterns on a polymeric surface with a feature size about one-fourth of the processing wavelength.

Figure 3.11 demonstrates the nanopatterning capabilities of DALP. In this experiment, three concentric square patterns were fabricated using a 100 μm droplet and 5 nJ femtosecond laser pulses, with pulse overlap providing continuous features of ~200 nm width at a shot-to-shot distance of 100 nm. The patterns are clearly visible under an optical microscope, both with the droplet present (figure 3.11(A)) and after droplet removal (figure 3.11(B)), confirming the robustness of the structures. Interestingly, imaging through the droplet magnifies the pattern by a factor of ~1.5, consistent with the refractive index of the liquid, and enhances resolution to the point that adjacent features become distinguishable only under droplet imaging. Atomic force microscopy characterization (figure 3.11(C)) reveals an average line width of ~350 nm and depth of ~50 nm, with narrow statistical deviations, while the slight lack of parallelism in the structure is attributed to stage positioning accuracy rather than limitations of the DALP method.

3.4.2 Role of liquid in pattern formation

Unlike ablation in air or vacuum, the presence of a liquid medium modifies optical feedback and energy dissipation [33, 36]. The surrounding liquid is not just a passive confining medium but an active director of nanoparticle self-organization and pattern formation. When the laser pulse ablates material from the target, the rapid heating and expansion of the plasma generate cavitation bubbles that trap atoms, clusters, and nanoparticles. As these bubbles expand and collapse, they create transient pressure and flow fields within the liquid. These hydrodynamic instabilities, combined with solvent viscosity and surface tension, dictate how nanoparticles are dispersed, aggregated, or aligned. For example, low-viscosity polar solvents allow rapid bubble oscillations that favor dispersed nanoparticles, while higher-viscosity or structured solvents promote chain-like or fractal assemblies due to slower diffusion and stronger interparticle interactions.

The chemical properties of the liquid further modulate nanoparticle patterning. Reactive solvents (e.g., water, alcohols, H_2O_2) generate radicals during plasma–liquid interactions that modify particle surfaces through oxidation or hydroxylation, influencing how nanoparticles interact and assemble. In contrast, nonpolar or less reactive liquids stabilize intermediate morphologies, such as nanowires, chains, or porous networks, by slowing down coalescence and recrystallization. Thus, by tuning solvent polarity, viscosity, and chemical reactivity, the liquid medium governs whether nanoparticles remain isolated, form linear chains, branched networks, or spherical aggregates. This makes the choice of liquid one of the most powerful tools for controlling nanoscale pattern formation in LAL without the need for external surfactants or templates. Moreover, liquids suppress debris redeposition, minimize oxidation, and in some cases, aid in localized chemical patterning through reactive byproducts.

It has been seen that the liquids with high refractive indices can promote interference effects, resulting in more defined LIPSS. On the other hand, solvents such as water, ethanol, or acetone exhibit different absorption and heat transfer behaviors, altering melt flow and surface tension effects.

3.4.3 Applications of nanopatterned surfaces via LAL

Laser-induced nanostructuring in liquids has been used in [37, 38]:

- **Photonic devices:** due to the creation of plasmonic surface gratings,
- **Superhydrophobic or wettability-tunable surfaces,**

- **Biosensors and antibacterial coatings** with textured metal oxide surfaces,
- **Catalyst supports** with increased surface area and active sites.

LAL thus bridges nanopatterning with simultaneous surface functionalization.

Bibliography

[1] Chen Q, Zhang C, Liu J, Ye Y, Li P and Liang C 2019 Solvent molecules dominated phase transition of amorphous Se colloids probed by in-situ Raman spectroscopy *Appl. Surf. Sci.* **466** 1000–6

[2] Zhang D, Gökce B and Barcikowski S 2017 Laser synthesis and processing of colloids: fundamentals and applications *Chem. Rev.* **117** 3990–4103

[3] Amendola V *et al* 2008 Room-temperature laser synthesis in liquid of oxide, metal-oxide core–shells, and doped oxide nanoparticles *J. Phys. Chem.* C **112** 8925–32

[4] Yang G W 2007 Laser ablation in liquids: applications in the synthesis of nanocrystals *Prog. Mater Sci.* **52** 648–98

[5] Zhang D, Li Z and Sugioka K 2021 Laser ablation in liquids for nanomaterial synthesis: diversities of targets and liquids *J. Phys. Photon.* **3** 042002

[6] Liu J, Liang C, Zhu X, Lin Y, Zhang H and Wu S 2016 Understanding the solvent molecules induced spontaneous growth of uncapped tellurium nanoparticles *Sci. Rep.* **6** 1–10

[7] Zeng H, Du X-W, Singh S C *et al* 2012 Laser-assisted fabrication of nanostructures for applications in plasmonics, photonics, and optoelectronics *Adv. Funct. Mater.* **22** 1333–53

[8] Amendola V and Meneghetti M 2009 Laser ablation synthesis in solution and size manipulation of noble metal nanoparticles *Phys. Chem. Chem. Phys.* **11** 3805–21

[9] Sylvestre J-P, Kabashin A V, Sacher E *et al* 2004 Surface chemistry of gold nanoparticles produced by laser ablation in aqueous media *J. Am. Chem. Soc.* **126** 7176–7

[10] Besner S and Meunier M 2010 Ultrafast dynamics of laser ablation in liquids: plasma formation and nanoparticle generation *J. Phys. Chem.* C **114** 10452–7

[11] Mafuné F, Kohno J Y, Takeda Y and Kondow T 2001 Formation and size reduction of gold nanoparticles by laser irradiation in aqueous solution *J. Phys. Chem.* B **105** 5114–20

[12] Dikovska A O, Avdeev M V, Evtukh A A *et al* 2017 Effect of laser pulse duration on the morphology and properties of nanoparticles generated by laser ablation of metals in liquids *Appl. Surf. Sci.* **418** 512–9

[13] Zhang C, Liu J, Ye Y, Aslam Z, Brydson R and Liang C 2018 Fe–N-doped mesoporous carbon with dual active sites loaded on reduced graphene oxides for efficient oxygen reduction catalysts *ACS Appl. Mater. Interfaces* **10** 2423–9

[14] Zhang C, Liu J, Ye Y, Chen Q and Liang C 2020 Encapsulation of Co-based nanoparticle in N-doped graphitic carbon for efficient oxygen reduction reaction *Carbon N. Y.* **156** 31–7

[15] Barcikowski S and Compagnini G 2013 Editorial: laser ablation in liquids: principles and applications in the preparation of nanomaterials *Phys. Chem. Chem. Phys.* **15** 3022–6

[16] Fazio E, Spadaro D, Corsaro C *et al* 2021 Laser ablation synthesis of oxide nanoparticles in water and ethanol: effects of environment and post-ablation aging *Nanomaterials* **11** 1681

[17] Sylvestre J-P, Poulin S, Kabashin A V *et al* 2005 Surface modification of gold nanoparticles produced by laser ablation in aqueous media *J. Phys. Chem.* B **109** 18299–306

[18] Semaltianos N G 2010 Nanoparticles by laser ablation *Appl. Phys.* A **101** 349–54

[19] Shih C-J, Lin H-C and Tai N-H 2012 Synthesis and characterization of iron oxide nanoparticles by laser ablation in liquid *Nanotechnology* **23** 125603
[20] Mizuno T, Nishimura T, Baba K *et al* 2018 Oxidation state control of Ce nanoparticles produced by laser ablation in liquid: effect of medium and laser fluence *J. Colloid Interface Sci.* **514** 262–8
[21] Liang C-H *et al* 2004 Preparation and characterization of gold–silver alloy nanoparticles by laser ablation in water *Langmuir* **20** 7283–7
[22] Tokumoto H, Yamamoto T A and Miyamoto Y 2003 Synthesis of gold nanoparticles by laser ablation in aqueous solution of surfactant *Mater. Lett.* **57** 3072–6
[23] Barcikowski S, Menéndez-Manjón A, Chichkov B N *et al* 2009 Generation and characterization of metal nanoparticles by femtosecond laser ablation in liquids *Appl. Phys.* A **94** 79–87
[24] Kohsakowski S, Streubel R *et al* 2019 Controlling colloid formation by femtosecond laser ablation in liquids through the additive-mediated reaction environment *Phys. Chem. Chem. Phys.* **21** 6378–85
[25] Kanitz A *et al* 2013 Impact of laser parameters and ambient environment on nanoparticle generation by laser ablation of metals in liquids *Phys. Chem. Chem. Phys.* **15** 3068–74
[26] Sajti C L *et al* 2004 Generation of nanoparticles by laser ablation of metal targets in liquids: formation mechanism and particle size distribution *J. Phys. Chem.* B **108** 5507–11
[27] Krawinkel J *et al* 2016 Electric field-assisted laser ablation in liquids for controlled nanoparticle generation *Appl. Phys.* A **122** 680
[28] Liu X *et al* 2010 Thermal effects in laser ablation of metal targets in liquid environments *Nanotechnology* **21** 265604
[29] Lam J *et al* 2015 Plasmon-enhanced photocatalysis using gold nanoparticles on TiO_2: impact of the spatial distribution *Adv. Funct. Mater.* **25** 405–18
[30] Dell'Aglio M *et al* 2015 Laser ablation in liquid and solid phase: a comparative study of plasma and nanoparticle characteristics *Spectrochim. Acta B At. Spectrosc.* **107** 1–7
[31] Dell'Aglio M, Gaudiuso R, De Pascale O and De Giacomo A 2015 Mechanisms and processes of pulsed laser ablation in liquids during nanoparticle production *Appl. Surf. Sci.* **348** 4–9
[32] Costache F, Kouteva-Arguirova S and Reif J 2003 Laser-induced periodic surface structures on bulk and thin film metals: influence of beam polarization and liquid environment *Appl. Surf. Sci.* **208–209** 486–91
[33] Ivanov D S *et al* 2021 Heat conduction and bubble formation in LAL-induced surface nanostructuring: insights from atomistic simulations *Appl. Phys.* A **127** 348
[34] Skoulas E, Tasolamprou A C, Kenanakis G and Stratakis E 2021 Laser induced periodic surface structures as polarizing optical elements *Appl. Surf. Sci.* **541** 148470
[35] Duocastella M, Florian C, Serra P and Diaspro A 2015 Sub-wavelength laser nanopatterning using droplet lenses *Sci. Rep.* **5** 16199
[36] Fuentes-Edfuf Y *et al* 2016 Laser surface nano/microstructuring of titanium in liquids: influence of scanning speed and medium *Appl. Surf. Sci.* **374** 81–91
[37] Yang K *et al* 2017 Nanopatterning of silicon surfaces via laser ablation in water: morphology control and photoluminescence properties *Appl. Surf. Sci.* **403** 427–34
[38] Lin Z *et al* 2019 Laser-induced self-organized surface structures in liquids: from nanoparticles to nanogratings *Opt. Lett.* **44** 3909–12

[39] Kim H J, Bang I C and Onoe J 2009 Characteristic stability of bare Au-water nanofluids fabricated by pulsed laser ablation in liquids *Opt. Lasers Eng.* **47** 532–8

[40] Scaramuzza S, Agnoli S and Amendola V 2015 Metastable alloy nanoparticles, metal-oxide nanocrescents and nanoshells generated by laser ablation in liquid solution: influence of the chemical environment on structure and composition *Phys. Chem. Chem. Phys.* **17** 28076–87

[41] Zhang D *et al* 2017 Formation mechanism of laser-synthesized iron-manganese alloy nanoparticles, manganese oxide nanosheets and nanofibers *Part. Part. Syst. Character.* **34** 1600225

Chapter 4

New progress in laser-based materials synthesis

In this chapter, we present the latest progress on nanomaterial synthesis using laser-based processing techniques. Recent advancements include the formation of high-entropy alloy nanoparticles, nanoparticle–polymer composites, doped semiconductor nanocrystals, and quantum dots, where the fundamental science and applications are significantly enhanced due to the innovations in laser-based methods [1–3]. The evolution of complex surface nanostructures and textures, which are formed via nanoscale mass transport under laser-induced thermal effects, is also included in this chapter.

4.1 High-entropy nanoparticles

The concept of high-entropy alloys (HEAs) is now ~21 years old from its inception in 2004 [4–6]. The idea was focused on the unexplored central regions of multielement phase diagrams, where all alloying elements are present in high concentrations, and there is no obvious base element. HEAs are originally defined as a blend of five or more elements with concentrations between 5 and 35 at.% [7]. Ni-based FCC HEAs [8–11], W-based BCC refractory HEAs [12, 13], Al-based low-density FCC HEAs [14], and noble-metal HEAs [15–18] are the primarily studied alloy systems showing single-phase solid-solutions. These systems, in general, have showcased several breakthroughs with their superior characteristics, including mechanical [19, 20], corrosion-resistant [21, 22], and radiation-resistant [23] properties. Recent studies also show that the field of HEA materials has since expanded to metallic glasses, metal oxides, carbides, and nitrides [24–26], and van der Waals materials (e.g., dichalcogenides, halides, and phosphorus trisulfide).

Motivated by the promising bulk properties of HEAs, the idea of downscaling into nanostructures with the incorporation of multiple immiscible elements into a single nanoparticle is considered an evolving research direction that has lately received a great amount of attention [27]. Despite the conceptual knowledge in bulk, early-stage development of HEA nanoparticles faced challenges in incorporating

doi:10.1088/978-0-7503-6224-5ch4

more than 2–3 elements without phase segregation [28]. While several prior studies demonstrated the formation of bi-/tri-/polyelemental nanoparticles [28–34] with phase segregation, the major advancements in this trajectory have recently been seen with the development of ultrafast nonequilibrium synthetic methodologies. These enabled a variety of HEA nanoparticles with homogeneous mixing even among immiscible elements. Yao *et al* in 2018 demonstrated the inclusion of multiple elements (Pt, Pd, Ni, Co, Fe, Au, Cu) using a carbothermal shock synthesis method, arguably the first study synthesizing HEA nanoparticles [16]. This method drives rapid heating and subsequent quenching of microsized precursors (2000 K in 55 ms), giving rise to multielemental mixing in resultant HEA nanoparticles. Subsequently, a few new methods have been developed based on the ultrafast kinetics-driven concept, which restricts elemental segregation. Rapid radiative heating or annealing [35–37], vapor phase spark discharge [38], acute chemical reduction [39, 40], low-temperature hydrogen spillover [41], sputtering [42–45], transient electrosynthesis [46], laser ablation in liquid (LAL) [47], plasma and microwave heating [48, 49] are the most noted thermal-shock methods governing in milliseconds timescale to enable forming HEA nanoparticles. A few notable examples are the PtPdFeCoNiAuCuSn, AgAuPtPdCu, and PtPdNiCoFe, which are proposed as nanocatalysts with significantly reduced content of precious metals [35, 50–52]. Studies have also reported on non-noble NiCoCr, NiCoCoFeMn, and $(Co_xMo_{0.7-x})Fe_{0.1}Ni_{0.1}Cu_{0.1}$ nanoparticles offering precious-metal-free electrocatalyst alternatives for ammonia oxidation, oxygen reduction reaction, and oxygen evolution reaction [42, 53, 54].

The application of lasers for synthesizing HEA nanoparticles has been primarily directed using the concept of (i) laser-induced dewetting of multielemental alloy films and (ii) LAL.

4.1.1 Nanosecond pulsed laser-induced dewetting

Following nonequilibrium pathways, nanostructure formation via nanosecond laser-induced dewetting (NLiD) presents promising opportunities due to its ultrafast ($\sim$100 ns), scalable, and cost-effective nature, with control on achieving a broad range of sizes, compositions, and structural morphologies [55, 56]. The driving force behind the formation of the nanoparticles is the breakup of energetically unstable thin films under laser irradiation, leading to the accumulation of material in a nanoscale droplet shape within nanoseconds. So far, nanoparticle formation of monometallic (Ag, Au, Ni, Co, Pt, etc) [56–59] and bimetallic (AgCo, AuCo, AuAg, CoCu, AgNi, etc) [31–33, 60–62] has been reported using the NLiD method. In addition, the formation of phase-segregated Janus and core–shell structures in bimetallic nanoparticles has also been demonstrated [33, 60]. While being investigated for producing a variety of monometallic and bimetallic nanoparticles, NLiD is a new technique to create MPEA nanoparticles with homogeneous mixing and unique microstructure reliably.

The first study on the formation of MPEA nanoparticles was recently reported by Sachan *et al* on NiCoCr nanoparticles, a well-known MPEA system, using ultrafast NLiD of alloy thin films [1]. The NiCoCr alloy shows a stark contrast in physical properties compared to individual metallic constituents, i.e., Ni, Co, and Cr, yet the transient nature of the laser-driven process facilitates a homogeneous distribution of the constituents (Ni, Co, and Cr) in the nanoparticles. Figure 4.1 shows a representative example of the progression of dewetting morphologies with an increasing number of laser pulse irradiations. As shown, nanoscale holes and polygonal structures are formed in the intermediate stages, leading to the final droplet-shaped NiCoCr nanoparticles forming using thin films of 15 and 30 nm.

It is also demonstrated that the contact angle of these droplet-shaped nanoparticles remains almost unchanged regardless of the nanoparticle size. In both cases, the average contact angle of nanoparticles is obtained at 145 ± 5. This shows the consistency in the elemental composition in the nanoparticles, which attributes to the invariable surface energies that are naturally independent of nanoparticle sizes. Figure 4.2 further shows the compositional mapping in the formed nanoparticles and their respective parent thin films to understand the elemental distribution and possible changes in composition, if any, during the process of dewetting. It is observed that the average atomic percentage of Ni, Cr, and Co are equilibrated at 42 ± 2 at.%, 39 ± 2 at.%, and 19 ± 2 at.%, respectively. It is seen that the composition of Co is almost half of that of Ni and Cr, which is caused by the intrinsic nature of laser ablation during the thin film deposition due to differences in the vapor pressures and optical properties of Ni, Co, and Cr [63]. The MPEA formation, based on the homogeneous elemental mixing, was validated on the nanoparticles formed on different substrates (SiO_2 and amorphous carbon), which is important as the formation of nanoparticles is highly dependent on the thermal transport that varies with the selection of different substrates. Via spectroscopic investigations, it is observed that NiCoCr nanoparticles are environmentally

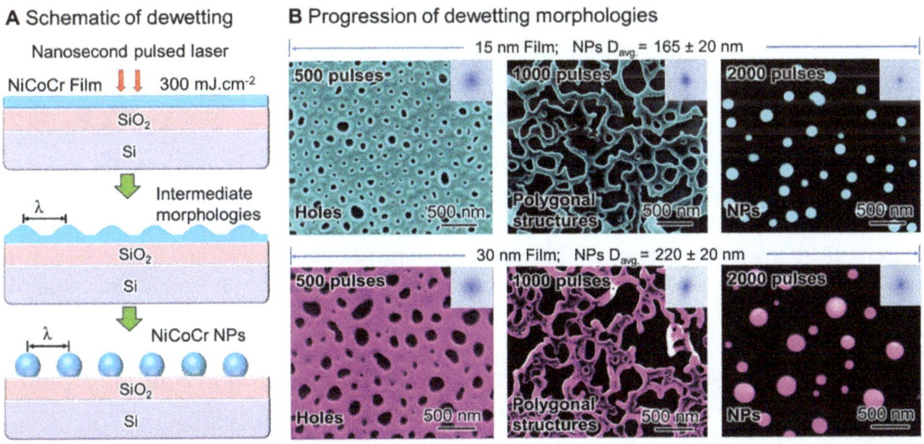

Figure 4.1. (a) The schematic diagram represents the NLiD process, (b) plan-view SEM images present the progression of dewetting morphologies with an increasing number of laser pulses showing holes, polygonal structures, and nanoparticles after 500, 1000, and 2000 pulses, respectively. The fast Fourier transform image in the inset shows the spatial short-range ordering of corresponding dewetting morphologies. [1] John Wiley & Sons. © 2024 Wiley-VCH GmbH.

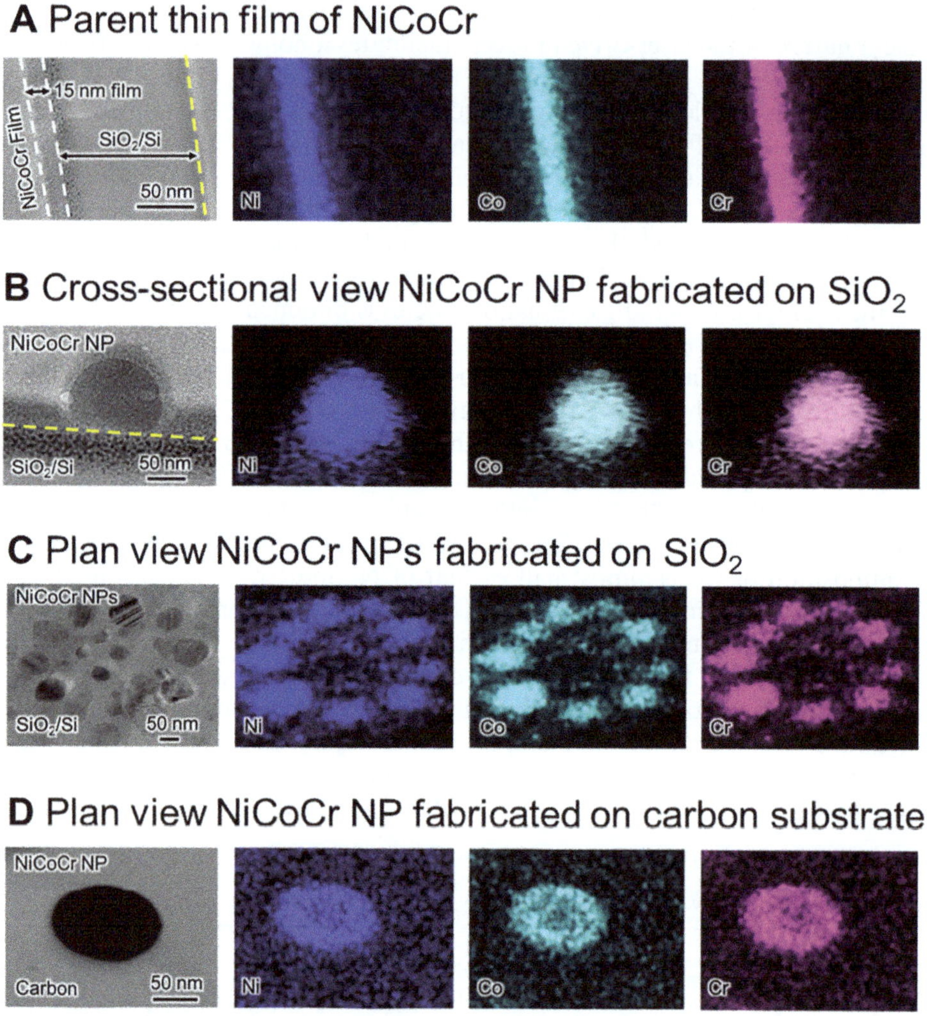

Figure 4.2. Cross-sectional view TEM image and elemental maps of Ni, Co and Cr of (A) NiCoCr parent thin films (~15 nm) and (B) formed a near-spherical nanoparticle on the SiO_2/Si substrate, (C) plan-view TEM image and the elemental distribution maps of Ni, Co, and Cr of NiCoCr nanoparticles fabricated on (C) SiO_2/Si substrate and (D) a carbon substrate. EDS maps of nanoparticles and thin film both show homogeneous elemental distribution of constituent Ni, Co, and Cr elements. [1] John Wiley & Sons. © 2024 Wiley-VCH GmbH.

stable and possibly seen with the enormous possibility of modifying their performance for prospective applications in catalysis and energy storage.

Recently, more follow-up work on synthesizing MPEA nanoparticles using nanosecond laser-induced dewetting has been reported by Liang *et al* [64] and Ghebretnsae *et al* [65]. Liang *et al* demonstrated the formation of spherical and polyhedral nanoparticles via kinetics-driven elemental trapping and nanoparticle restructuring through thermodynamics [64]. The work by Ghebretnsae *et al* showed

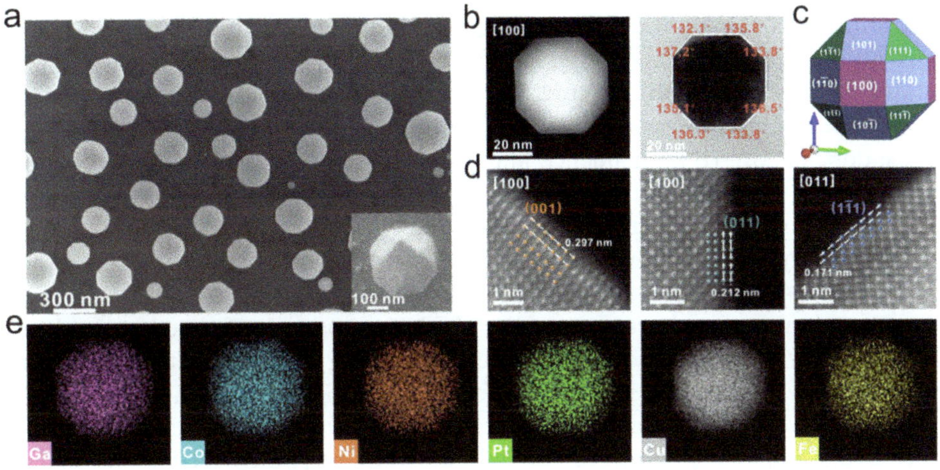

Figure 4.3. Morphology, structure, and elemental characterization of HEA polyhedra. (a) SEM image of GaPtFeCoNiCu polyhedra. (b) The HAADF-STEM image (left) and TEM image (right) of an individual GaPtFeCoNiCu polyhedron. The angles between surface planes were plotted in the images, where the white lines were drawn to highlight the edge as a guide to the eye. (c) An ideal three-dimensional model of the HEA polyhedron surrounded by {100}, {110}, and {111} facets. (d) The atomic-resolution HAADF-STEM images of the edges from [100] and [011] directions. The colored spheres and white lines were placed to highlight the atoms and crystal faces, respectively. (e) The EDS elemental mappings of the particle. Reprinted with permission from [64]. Copyright (2025) American Chemical Society.

the dewetting of multilayers of noble metal in order to form HEA nanoparticles [65]. Figure 4.3 shows the morphological, structural, and elemental characterization of HEA polyhedral of GaPtFeCoNiCu nanoparticles.

4.1.2 Laser ablation in liquid

Among the various synthesis approaches, LAL has garnered attention due to its chemical-free, surfactant-free, and scalable route to produce high-purity colloidal nanoparticles. LAL enables the formation of HEA nanoparticles in a single-step, bottom-up process, driven by extreme nonequilibrium conditions. LAL involves focusing high-intensity laser pulses (typically ns or fs duration) onto a solid HEA target submerged in a liquid medium (e.g., water, ethanol). The laser–target interaction leads to rapid energy deposition, surface melting, plasma formation, and material ejection into the surrounding liquid. The ablated atoms and clusters undergo ultrafast cooling, nucleation, and growth into nanoparticles. The key phenomena in LAL include:

(i) plasma plume formation and cavitation bubble dynamics;
(ii) nonequilibrium nucleation and growth during rapid quenching ($\sim 10^9$ K s^{-1});
(iii) laser-induced mixing and alloying of multi-elemental species;
(iv) stabilization of nanoparticles by the surrounding liquid environment.

The absence of ligands or chemical precursors allows the formation of ligand-free nanoparticles, making LAL an ideal method for catalysis-related applications.

In the context of HEAs, the multi-elemental target composition allows for simultaneous ablation of diverse atomic species. The high-temperature plasma plume contains a supersaturated mixture of elements that rapidly nucleates to form nanoparticles. The success of forming a homogeneous single-phase HEA nanoparticle depends on: (i) laser fluence and pulse duration; (ii) elemental vaporization thresholds and relative ablation rates; (iii) cooling rate and diffusivity of constituent atoms in the plume; and (iv) liquid environment, which affects bubble collapse, particle stability, and oxidation. Figure 4.4 depicts the laser-based synthesis of HEA nanoparticles [47].

Experimental studies have shown that HEA nanoparticles formed by LAL often exhibit FCC or BCC solid solution phases, depending on the target composition. A recent study by Waag *et al* showed the formation of HEA nanoparticles consisting of NiCoCrFeMn, which is a known HEA, also known as Cantor alloy. This study reported the single-step synthesis of colloidal NiCoCrFeMn HEA nanoparticles with targeted equimolar stoichiometry and diameters less than 5 nm by liquid-phase, ultrashort-pulsed laser ablation of the consolidated and heat-treated micropowders of the five constituent metals.

HRTEM and EDS confirm uniform elemental distribution, while x-ray diffraction (XRD) shows broadened peaks indicating nanoscale crystallinity of the formed nanoparticles. As shown in figure 4.5, the composition of constituent elements in the ablation target and in ablated HEA nanoparticles was consistent. While there is a possibility of having compositional changes from one nanoparticle to the other, a homogeneous chemical composition of single ultrasmall nanoparticles was verified based on the crystal structure analysis. No strong deviations from the chemical

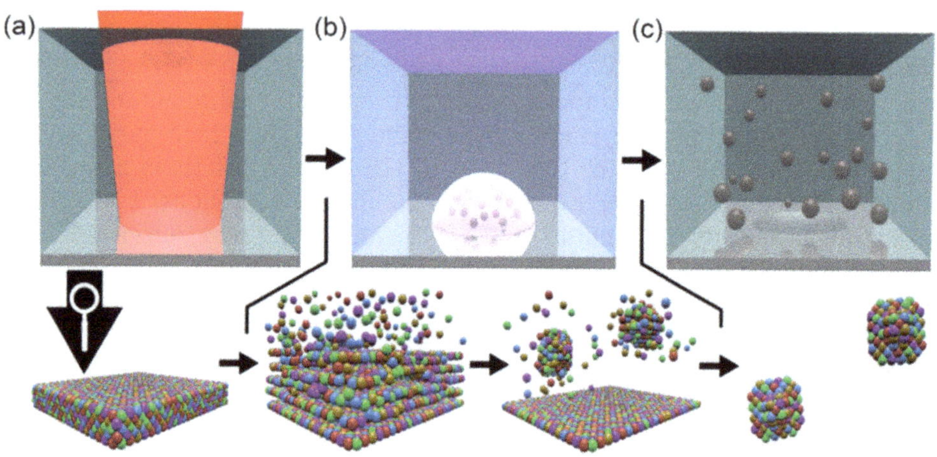

Figure 4.4. Qualitative representation of the laser-based synthesis of HEA nanoparticles. The synthesis method consisted of the following stages: ultrashort-pulsed laser irradiation of the bulk HEA (a), the atomization/ionization of the bulk causing the formation of a plume, and subsequent nucleation and condensation of the ablated matter in the vapor phase of the liquid (b) and the colloidal HEA nanoparticles electrostatically stabilized in ethanol (c). Reproduced from [47] with permission from the Royal Society of Chemistry.

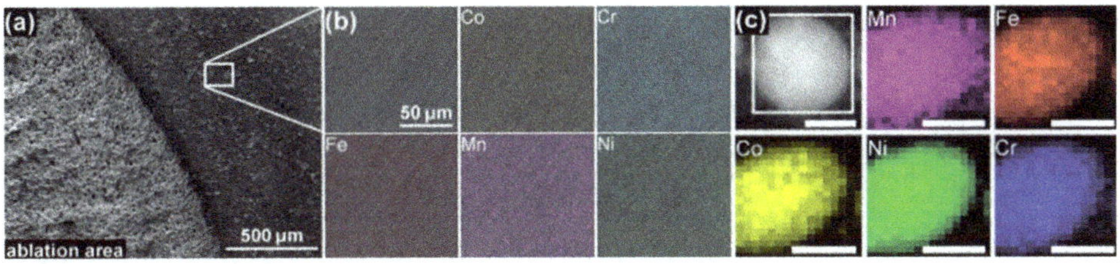

Figure 4.5. Scanning electron microscope image (secondary electrons) of an ablation target near the border of an ablation area (a), elemental maps of Co, Cr, Fe, Mn, and Ni obtained by energy dispersive x-ray spectroscopy of a nonablated target area (b), and scanning transmission electron microscopy image and elemental maps of Co, Cr, Fe, Mn, and Ni for a single nanoparticle (c). In (a), the laser ablation area is located on the left side of the image. In (c), all scale bars represent 25 nm. Reproduced from [47] with permission from the Royal Society of Chemistry.

composition of the ablation target are observed either by the selected area diffraction or XRD analysis.

4.2 Nanoparticle–polymer composites

Nanoparticle–polymer composites, also known as nanohybrids or hybrid composites, are gaining considerable interest due to their tunable properties and broad application potential. These materials combine the functional advantages of both nanoparticles and polymers, enabling applications ranging from mechanical reinforcement and flame retardancy to antimicrobial coatings, drug delivery systems, photovoltaics, fuel cells, and sensors [66–68]. Synthesis methods for these composites are generally classified into physical and chemical approaches, depending on the nature of interactions between the nanoparticles and the polymer matrix [69]. Fabrication is usually achieved either by incorporating nanoparticles into a molten polymer followed by extrusion, or by dispersing them in a polymer solution with subsequent solvent evaporation. The melt-blending method is especially effective for generating composite fibers through electrospinning or creating three-dimensional (3D) structures using injection molding, whereas the solution-based strategy is more commonly applied for producing thin films by casting [66, 70–72]. Physical methods are especially favored for large-scale production of bulk composites due to their operational simplicity, cost-efficiency, and scalability.

4.3 Laser-synthesis of colloidal particles

Laser-based synthesis offers a clean, surfactant-free, and highly controllable route to produce colloidal nanoparticles, which can be readily integrated into polymer matrices to form functional composites. This chapter explores the emerging strategy of utilizing pulsed LAL (PLAL) to generate metallic and metal oxide nanoparticles directly in polymer-containing solutions or subsequently dispersing them into polymer matrices. The resulting colloidal nanoparticle–polymer composites exhibit unique synergies, combining the tunable physicochemical properties of laser-generated nanoparticles, such as size, surface energy, and crystallinity, with the flexibility, processability, and functionality of polymers.

The laser synthesis approach eliminates the need for chemical precursors or reducing agents, allowing for the formation of high-purity nanoparticles that are ideal for biomedical, optoelectronic, and sensing applications. Moreover, *in situ* laser ablation within polymer solutions facilitates direct encapsulation or surface functionalization of nanoparticles, promoting strong interfacial interactions and enhanced stability. The physicochemical characteristics of the composites can be tailored by varying laser parameters (wavelength, pulse duration, fluence), target material, and polymer type or concentration. When adopting LAL synthesis routes, polymer concentration plays an important role in the resulting physical properties (e.g., size and stability) of the synthesized particle-polymer composites. For example, the size of *in situ* LAL-synthesized encapsulated nanoparticle-polymer composites is first reduced as the polymer concentration increases and becomes saturated at a certain polymer concentration (e.g., FeO@PVP composites, figure 4.6 (a–d) [66]. In addition, polymer concentration (typically a few wt%) also influences the particle dispersion and mass productivity. Wagener *et al* showed that 0.3 wt% polyurethane (figure 6e–h) is an optimal concentration to achieve well-dispersed Ag-TPU composites with 9 mg h^{-1}, while using lower and higher polymer concentrations with respect to this value will result in composite agglomeration and reduced productivity, respectively [73]. When alloying two types of nanoparticles or fragmenting polymer-capped nanoparticles, polymer concentration-dependent effects must also be considered. For example, Izgaliev *et al* reported that the rate of alloying behavior depends on the PVP concentration and the addition of PVP improved the stability of alloyed AgAu nanoparticles [74]. However, as the PVP

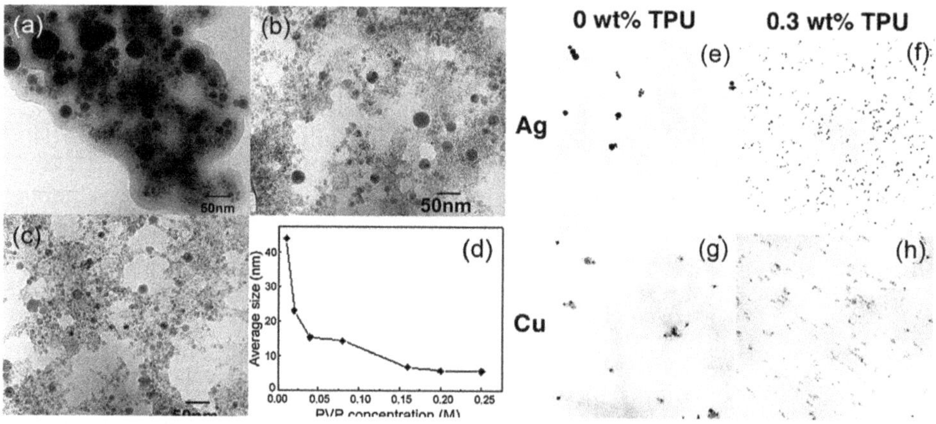

Figure 4.6. (a–c) TEM images of FeO@PVP composites synthesized by *in situ* LAL of an iron plate with different PVP concentrations of 0.01 M, 0.02 M, and 0.08 M, respectively. (d) The average size of FeO@PVP composites as a function of PVP concentration. (a–d) Reprinted with permission from [76]. Copyright (2008) American Chemical Society. (e–h) Electron micrographs of ultra-thin films of nanocomposites generated with different TPU concentrations during laser ablation ((e–f): silver, (g–h): copper). (e–h) Reproduced from [73] with permission from the Royal Society of Chemistry.

concentration increases, a capping effect prevents alloying between Au nanoparticles and Ag nanoparticles, and eventually the alloying behavior is terminated when the PVP concentration exceeds 0.5 g l^{-1} [74, 75].

This synthesis methodology presents a green and scalable platform for developing next-generation nanocomposites with tailored properties, paving the way for applications in drug delivery, antimicrobial coatings, flexible electronics, and environmental remediation.

4.4 Doped semiconductor nanocrystals

Laser-based techniques have emerged as powerful tools for synthesizing doped semiconductor nanocrystals (NCs) with controlled composition, crystallinity, and doping profiles. In particular, PLAL, laser fragmentation, and laser irradiation of precursors in colloidal or solid-phase systems offer surfactant-free, high-purity synthesis routes for doped nanocrystals such as ZnO:Mn, Si:Er, or CdSe:Cu.

The high energy density and ultrafast timescales of laser pulses enable non-equilibrium incorporation of dopants into semiconductor matrices, overcoming limitations of conventional thermal doping processes. This laser-induced doping often results in unique optoelectronic and magnetic properties arising from quantum confinement effects, localized states, and lattice strain induced by the dopant atoms.

Critical parameters such as laser fluence, wavelength, pulse duration, and ambient medium significantly influence doping efficiency, dopant distribution (substitutional versus interstitial), and nanocrystal size. Furthermore, laser processing allows for precise spatial and temporal control, enabling post-synthesis tuning of doping levels or bandgap modulation through laser annealing or defect engineering.

These laser-induced doped NCs have demonstrated promise in applications such as light-emitting diodes, photocatalysis, spintronics, and bioimaging. The scalability, versatility, and clean synthesis environment make laser-based doping a frontier technique in nanomaterials engineering.

4.5 Quantum dots

Quantum dots (QDs) are semiconductor NCs that exhibit unique size-dependent electronic and optical properties due to quantum confinement, making them attractive for applications in bioimaging, optoelectronics, photovoltaics, and quantum information technologies [77, 78]. Traditional chemical synthesis methods often require toxic solvents, stabilizing ligands, and multistep processes that can limit biocompatibility and surface purity. In contrast, laser-induced synthesis, particularly PLAL, offers a surfactant-free, one-step, and environmentally friendly route to produce high-quality QDs with clean surfaces [79, 80].

In PLAL, high-energy laser pulses are focused onto a solid target submerged in a liquid medium, generating a transient plasma plume. Rapid quenching leads to nucleation and growth of nanoparticles, including QDs, directly in the liquid phase. Materials such as CdSe, ZnO, Si, MoS_2, and graphitic carbon have been successfully used to generate QDs with tunable size and photoluminescence [81].

Laser parameters such as fluence, pulse duration, repetition rate, and wavelength, along with liquid medium composition, can be precisely adjusted to control QD size, crystallinity, and surface state, enabling emission tunability across the UV to near-IR spectrum [82]. Furthermore, laser fragmentation of pre-formed nanoparticles can refine size distributions and improve optical uniformity [83, 84].

Key advantages of laser-induced QD synthesis include:

- surfactant-free and contamination-free surfaces, suitable for biomedical and optical applications,
- rapid, scalable, and solvent-compatible processing, enabling high-purity production,
- *in situ* doping and surface functionalization, using dopant precursors in solution or co-ablation with dopant-containing targets,
- post-synthesis laser annealing or irradiation, enabling further tuning of emission and crystallinity.

Recent studies demonstrate the successful generation of graphene QDs with blue to green fluorescence, ZnO QDs for UV photodetectors, and Si QDs for biosensing and LED applications. These QDs show competitive or superior performance to chemically synthesized analogs, without requiring post-synthetic purification.

In summary, laser-induced synthesis of QDs presents a powerful, clean, and tunable platform for creating high-quality quantum nanomaterials. The method aligns with sustainable nanomanufacturing goals and offers compatibility with biomedical and optoelectronic device integration.

Bibliography

[1] Mandal S, Gupta A K, Konečná A, Shirato N, Hachtel J A and Sachan R 2024 Creation of multi-principal element alloy NiCoCr nanostructures via nanosecond laser-induced dewetting *Small* **20** 2309574

[2] Xiao J, Liu P, Wang C X and Yang G W 2017 External field-assisted laser ablation in liquid: an efficient strategy for nanocrystal synthesis and nanostructure assembly *Prog. Mater Sci.* **87** 140–220

[3] Fu M, Ma X, Zhao K, Li X and Su D 2021 High-entropy materials for energy-related applications *iScience* **24** 102177

[4] Cantor B, Chang I T H, Knight P and Vincent A J B 2004 Microstructural development in equiatomic multicomponent alloys *Mater. Sci. Eng.* A **375–7** 213–8

[5] Yeh J W, Chen S K, Lin S J, Gan J Y, Chin T S, Shun T T, Tsau C H and Chang S Y 2004 Nanostructured high-entropy alloys with multiple principal elements: novel alloy design concepts and outcomes *Adv. Eng. Mater.* **6** 299–303

[6] Miracle D B 2019 High entropy alloys as a bold step forward in alloy development *Nat. Commun.* **10** 1–3

[7] Gorsse S, Couzinié J and Miracle D B 2018 From high-entropy alloys to complex concentrated alloys *C. R. Phys.* **19** 721–36

[8] Zhang Y *et al* 2015 Influence of chemical disorder on energy dissipation and defect evolution in concentrated solid solution alloys *Nat. Commun.* **6** 8736

[9] Lu C *et al* 2016 Enhancing radiation tolerance by controlling defect mobility and migration pathways in multicomponent single phase alloys *Nat. Commun.* 1–8
[10] An Z, Jia H, Wu Y, Rack P D, Patchen A D, Liu Y, Ren Y, Li N and Liaw P K 2015 Solid-solution CrCoCuFeNi high-entropy alloy thin films synthesized by sputter deposition *Mater. Res. Lett.* **3** 203–9
[11] He M R, Wang S, Shi S, Jin K, Bei H, Yasuda K, Matsumura S, Higashida K and Robertson I M 2017 Mechanisms of radiation-induced segregation in CrFeCoNi-based single-phase concentrated solid solution alloys *Acta Mater.* **126** 182–93
[12] El-Atwani O *et al* 2019 Outstanding radiation resistance of tungsten-based high-entropy alloys *Sci. Adv.* **5** 1–28
[13] Tunes M A and Vishnyakov V M 2019 Microstructural origins of the high mechanical damage tolerance of NbTaMoW refractory high-entropy alloy thin films *Mater. Des.* **170** 107692
[14] Koch C C 2017 Nanocrystalline high-entropy alloys *J. Mater. Res.* **32** 3435–44
[15] Nellaiappan S, Katiyar N K, Kumar R, Parui A, Malviya K D, Pradeep K G, Singh A K, Sharma S, Tiwary C S and Biswas K 2020 High-entropy alloys as catalysts for the CO_2 and CO reduction reactions: experimental realization *ACS Catal.* **10** 3658–63
[16] Yao Y *et al* 2018 Carbothermal shock synthesis of high-entropy-alloy nanoparticles *Science (1979)* **359** 1489–94
[17] Lacey S D *et al* 2019 Stable multimetallic nanoparticles for oxygen electrocatalysis *Nano Lett.* **19** 5149–58
[18] Song B *et al* 2020 In situ oxidation studies of high-entropy alloy nanoparticles *ACS Nano* **14** 15131–43
[19] Jiang W, Zhu Y and Zhao Y 2022 Mechanical properties and deformation mechanisms of heterostructured high-entropy and medium-entropy alloys: a review **8** 1–17
[20] Uzer B *et al* 2018 On the mechanical response and microstructure evolution of NiCoCr single crystalline medium entropy alloys *Mater. Res. Lett.* **3831** 442–9
[21] Yin B, Yoshida S, Tsuji N and Curtin W A 2020 Yield strength and misfit volumes of NiCoCr and implications for short-range-order *Nat. Commun.* **11** 1–7
[22] Xu D, Wang M, Li T, Wei X and Lu Y 2022 A critical review of the mechanical properties of CoCrNi-based medium-entropy alloys *Microstruct.* **2** 2022001
[23] Sachan R, Ullah M W, Chisholm M F, Weber W J, Liu J, Zhai P, Kluth P, Trautmann C, Bei H and Zhang Y 2018 Extreme elastic and inelastic interactions in concentrated solid solutions *Mater. Des.* **150** 1–8
[24] Gild J, Samiee M, Braun J L, Harrington T, Vega H, Hopkins P E, Vecchio K and Luo J 2018 High-entropy fluorite oxides *J. Eur. Ceram. Soc.* **38** 3578–84
[25] Sharma Y *et al* 2018 Single crystal high entropy perovskite oxide epitaxial films *Phys. Rev. Mater.* **2** 060404
[26] Li F, Zhou L, Liu J X, Liang Y and Zhang G J 2019 High-entropy pyrochlores with low thermal conductivity for thermal barrier coating materials *J. Adv. Ceram.* **8** 576–82
[27] Yao Y *et al* 2022 High-entropy nanoparticles: synthesis-structureproperty relationships and data-driven discovery *Science (1979)* **376**
[28] Chen P-C, Liu X, Hedrick J L, Xie Z, Wang S, Lin Q-Y, Hersam M C, Dravid V P and Mirkin C A 2016 Polyelemental nanoparticle libraries *Science (1979)* **352** 1565
[29] Chen P C, Liu M, Du J S, Meckes B, Wang S, Lin H, Dravid V P, Wolverton C and Mirkin C A 2019 Interface and heterostructure design in polyelemental nanoparticles *Science (1979)* **363** 959–64

[30] Huynh K, Pham X, Kim J, Lee S H, Chang H, Rho W and Jun B 2020 Synthesis, properties, and biological applications of metallic alloy nanoparticles *Int. J. Mol. Sci.* **21** 5174

[31] Sachan R, Yadavali S, Shirato N, Krishna H, Ramos V, Duscher G, Pennycook S J, Gangopadhyay A K, Garcia H and Kalyanaraman R 2012 Self-organized bimetallic Ag–Co nanoparticles with tunable localized surface plasmons showing high environmental stability and sensitivity *Nanotechnology* **23** 275604

[32] Sachan R, Ramos V, Malasi A, Yadavali S, Bartley B, Garcia H, Duscher G and Kalyanaraman R 2013 Oxidation-resistant silver nanostructures for ultrastable plasmonic applications *Adv. Mater.* **25** 2045–50

[33] Sachan R, Malasi A, Ge J, Yadavali S, Krishna H, Gangopadhyay A, Garcia H, Duscher G and Kalyanaraman R 2014 Ferroplasmons: intense localized surface plasmons in metal-ferromagnetic nanoparticles *ACS Nano* **8** 9790–8

[34] Kwon S G *et al* 2015 Heterogeneous nucleation and shape transformation of multi-component metallic nanostructures *Nat. Mater.* **14** 215–23

[35] Gao S, Hao S, Huang Z, Zhang X and Shahbazian-yassar R 2020 Synthesis of high-entropy alloy nanoparticles on supports by the fast moving bed pyrolysis *Nat. Commun.* **11** 2016

[36] Yang Y, Song B, Ke X, Xu F, Bozhilov K N, Hu L, Shahbazian-yassar R and Zachariah M R 2020 Aerosol synthesis of high entropy alloy nanoparticles *Langmuir* **36** 1985–92

[37] Chen Y, Zhan X, Bueno S L A, Shafei I H, Ashberry H M, Chatterjee K, Xu L, Tang Y and Skrabalak S E 2021 Synthesis of monodisperse high entropy alloy nanocatalysts from core@shell nanoparticles *Nanoscale Horiz.* **6** 231–7

[38] George E P, Raabe D and Ritchie R O 2019 High-entropy alloys *Nat. Rev. Mater.* **4** 515–34

[39] Wu D, Kusada K, Yamamoto T, Toriyama T, Matsumura S, Kawaguchi S, Kubota Y and Kitagawa H 2020 Platinum-group-metal high-entropy-alloy nanoparticles *J. Am. Chem. Soc.* **142** 13833–8

[40] Kusada K, Wu D and Kitagawa H 2020 New aspects of platinum group metal-based solid-solution alloy nanoparticles: binary to high-entropy alloys *Chem. Eur. J.* **26** 5105–30

[41] Mori K, Hashimoto N, Kamiuchi N, Yoshida H, Kobayashi H and Yamashita H 2021 Hydrogen spillover-driven synthesis of high-entropy alloy nanoparticles as a robust catalyst for CO_2 hydrogenation *Nat. Commun.* **12** 1–11

[42] Löffler T, Meyer H, Savan A, Wilde P, Garzón Manjón A, Chen Y T, Ventosa E, Scheu C, Ludwig A and Schuhmann W 2018 Discovery of a multinary noble metal–free oxygen reduction catalyst *Adv. Energy Mater.* **8** 1802269

[43] Löffler T, Savan A, Meyer H, Meischein M, Strotkötter V, Ludwig A and Schuhmann W 2020 Design of complex solid-solution electrocatalysts by correlating configuration, adsorption energy distribution patterns, and activity curves *Angew.Chem.Int. Ed.* **59** 5844–50

[44] Batchelor T A A *et al* 2021 Complex-solid-solution electrocatalyst discovery by computational prediction and high-throughput experimentation *Angew. Chem. Int. Ed.* **60** 6932–7

[45] Ludwig A 2019 Discovery of new materials using combinatorial synthesis and high-throughput characterization of thin-film materials libraries combined with computational methods *NPJ Comput. Mater.* **5** 70

[46] Glasscott M W, Pendergast A D, Goines S, Bishop A R, Hoang A T, Renault C and Dick J E 2019 Electrosynthesis of high-entropy metallic glass nanoparticles for designer, multi-functional electrocatalysis *Nat. Commun.* **10** 2650

[47] Waag F, Li Y, Ziefuß A R, Bertin E, Kamp M, Duppel V, Marzun G, Kienle L, Barcikowski S and Gökce B 2019 Kinetically-controlled laser-synthesis of colloidal high-entropy alloy nanoparticles *RSC Adv.* **9** 18547–58

[48] Qiao H *et al* 2021 Scalable synthesis of high entropy alloy nanoparticles by microwave heating *ACS Nano* **15** 14928–37

[49] Yao Y, Dong Q and Hu L 2019 Overcoming immiscibility via a milliseconds-long 'Shock' synthesis toward alloyed nanoparticles *Matter* **1** 1451–3

[50] Xin Y, Li S, Qian Y, Zhu W, Yuan H, Jiang P, Guo R and Wang L 2020 High-entropy alloys as a platform for catalysis: progress, challenges, and opportunities *ACS Catal.* **10** 11280–306

[51] Yao Y *et al* 2021 Extreme mixing in nanoscale transition metal alloys *Matter* **4** 2340–53

[52] Wang X *et al* 2020 Continuous synthesis of hollow high-entropy nanoparticles for energy and catalysis applications *Adv. Mater.* **32** 2002853

[53] Yao Y, Dong Q, Brozena A, Luo J, Miao J, Chi M, Anapolsky A and Hu L 2022 High-entropy nanoparticles: synthesis-structure- property relationships and data-driven discovery *Science (1979)* **376** 1–11

[54] Tomboc G M, Kwon T, Joo J and Lee K 2020 High entropy alloy electrocatalysts: a critical assessment of fabrication and performance *J. Mater. Chem. A Mater.* **8** 14844–62

[55] Trice J, Thomas D, Favazza C, Sureshkumar R and Kalyanaraman R 2007 Pulsed-laser-induced dewetting in nanoscopic metal films: theory and experiments *Phys. Rev. B Condens. Matter Mater. Phys.* **75** 1–15

[56] Krishna H, Sachan R, Strader J, Favazza C, Khenner M and Kalyanaraman R 2010 Thickness-dependent spontaneous dewetting morphology of ultrathin Ag films *Nanotechnology* **21** 155601

[57] Krishna H, Shirato N, Favazza C and Kalyanaraman R 2009 Energy driven self-organization in nanoscale metallic liquid films *Phys. Chem. Chem. Phys.* **11** 8136–43

[58] Favazza C, Trice J, Krishna H, Kalyanaraman R and Sureshkumar R 2006 Laser-induced short- and long-range orderings of Co nanoparticles on SiO_2 *Appl. Phys. Lett.* **88** 2004–7

[59] Yadavali S, Khenner M and Kalyanaraman R 2013 Pulsed laser dewetting of Au films: experiments and modeling of nanoscale behavior *J. Mater. Res.* **28** 1715–23

[60] McKeown J T, Wu Y, Fowlkes J D, Rack P D and Campbell G H 2015 Simultaneous in-situ synthesis and characterization of co@cu core-shell nanoparticle arrays *Adv. Mater.* **27** 1060–5

[61] Fowlkes J D, Wu Y and Rack P D 2010 Directed assembly of bimetallic nanoparticles by pulsed-laser-induced dewetting: a unique time and length scale regime *ACS Appl. Mater. Interfaces* **2** 2153–61

[62] Garfinkel D A, Tang N, Pakeltis G, Emery R, Ivanov I N, Gilbert D A and Rack P D 2022 Magnetic and optical properties of Au-Co solid solution and phase-separated thin films and nanoparticles *ACS Appl. Mater. Interfaces* **14** 15047–58

[63] Mandal S, Kumar A, Rose V, Wieghold S, Shirato N and Sachan R 2023 Understanding non-stochiometric deposition of multi-principal elemental NiCoCr thin films *Appl. Surf. Sci.* **623** 157011

[64] Liang J *et al* 2025 Synthesis of high-entropy alloy polyhedra using liquid metal dewetting *J. Am. Chem. Soc.* **147** 16742–6

[65] Ghebretnsae S, Ziehl T J, Purdy S, Zhang P, Sham T K and Shi Y 2025 Pulsed laser-induced dewetting for the production of noble-metal high-entropy-alloy nanoparticles *Nanoscale* **17** 15423–35

[66] Zhang D and Gökce B 2017 Perspective of laser-prototyping nanoparticle-polymer composites *Appl. Surf. Sci.* **392** 991–1003

[67] Tripathi B P and Shahi V K 2011 Organic–inorganic nanocomposite polymer electrolyte membranes for fuel cell applications *Prog. Polym. Sci.* **36** 945–79

[68] Kickelbick G 2003 Concepts for the incorporation of inorganic building blocks into organic polymers on a nanoscale *Prog. Polym. Sci.* **28** 83–114

[69] Li S, Meng Lin M, Toprak M S, Kim D K and Muhammed M 2010 Nanocomposites of polymer and inorganic nanoparticles for optical and magnetic applications *Nano Rev.* **1** 5214

[70] Schwenke A, Wagener P, Weiß A, Klimenta K, Wiegel H, Sajti L and Barcikowski S 2013 Laserbasierte Generierung matrixbinderfreier Nanopartikel-Polymerkomposite für bioaktive Medizinprodukte *Chem. Ing. Tech.* **85** 740–6

[71] Shih C Y, Chen C, Rehbock C, Tymoczko A, Wiedwald U, Kamp M, Schuermann U, Kienle L, Barcikowski S and Zhigilei L V 2021 Limited elemental mixing in nanoparticles generated by ultrashort pulse laser ablation of AgCu bilayer thin films in a liquid environment: atomistic modeling and experiments *J. Phys. Chem.* C **125** 2132–55

[72] Wagener P, Jakobi J, Rehbock C, Chakravadhanula V S K, Thede C, Wiedwald U, Bartsch M, Kienle L and Barcikowski S 2016 Solvent-surface interactions control the phase structure in laser-generated iron-gold core-shell nanoparticles *Sci. Rep.* **6** 1–12

[73] Wagener P, Brandes G, Schwenke A and Barcikowski S 2011 Impact of in situpolymer coating on particle dispersion into solid laser-generated nanocomposites *Phys. Chem. Chem. Phys.* **13** 5120–6

[74] Izgaliev A T, Simakin A V, Shafeev G A and Bozon-Verduraz F 2004 Intermediate phase upon alloying Au-Ag nanoparticles under laser exposure of the mixture of individual colloids *Chem. Phys. Lett.* **390** 467–71

[75] Izgaliev A T, Simakin A V and Shafeev G A 2004 Formation of the alloy of Au and Ag nanoparticles upon laser irradiation of the mixture of their colloidal solutions *Kvantovaya Elektronika* **34** 47–51

[76] Liu P, Cai W and Zeng H 2008 Fabrication and size-dependent optical properties of FeO nanoparticles induced by laser ablation in a liquid medium *J. Phys. Chem.* C **112** 3261–6

[77] Alivisatos A P 1996 Semiconductor clusters, nanocrystals, and quantum dots *Science (1979)* **271** 933–7

[78] Michalet X, Pinaud F F, Bentolila L A, Tsay J M, Doose S, Li J J, Sundaresan G, Wu A M, Gambhir S S and Weiss S 2005 Quantum dots for live cells, *in vivo* Imaging, and diagnostics *Science (1979)* **307** 538–44

[79] Lévy A *et al* 2021 Surface chemistry of gold nanoparticles produced by laser ablation in pure and saline water *Langmuir* **37** 5783–94

[80] Zeng H, Du X-W, Singh S C, Kulinich S A, Yang S, He J and Cai W 2012 Nanomaterials via laser ablation/irradiation in liquid: a review *Adv. Funct. Mater.* **22** 1333–53

[81] Dong Y, Shao J, Chen C, Li H, Wang R, Chi Y, Lin X and Chen G 2012 Blue luminescent graphene quantum dots and graphene oxide prepared by tuning the carbonization degree of citric acid *Carbon N. Y.* **50** 4738–43

[82] Besner S, Kabashin A V and Meunier M 2006 Fragmentation of colloidal nanoparticles by femtosecond laser-induced supercontinuum generation *Appl. Phys. Lett.* **89** 233122

[83] Huisken F, Hofmeister H, Kohn B, Laguna M A and Paillard V 2000 Laser production and deposition of light-emitting silicon nanoparticles *Appl. Surf. Sci.* **154–155** 305–13

[84] Zabotnov S V *et al* 2020 Nanoparticles produced via laser ablation of porous silicon and silicon nanowires for optical bioimaging *Sensors (Switzerland)* **20** 1–16

IOP Publishing

Laser-assisted Formation of Metallic Nanoparticles
Theory, fabrication and applications
Ritesh Sachan

Chapter 5

Microstructure evolution and characterization: a microscopy study of laser-synthesized nanoparticles

The evolution of microstructure in laser-synthesized nanoparticles is central to understanding their unique physicochemical properties and technological potential. This chapter presents a comprehensive microscopy-driven study of nanoparticles produced through pulsed-laser-based synthesis routes, including pulsed laser deposition, pulsed laser-induced dewetting, and laser ablation in liquids (LAL). We highlight how advanced electron and scanning probe microscopies, such as high-resolution x-ray diffraction, SEM, TEM, EDS, electron energy-loss spectroscopy, and four-dimensional scanning transmission electron microscopy, reveal nanoscale features ranging from crystallinity, defects, and compositional inhomogeneities to morphology, interfaces, and self-assembly pathways. Special emphasis is placed on correlating synthesis parameters with microstructural evolution, including grain refinement, surface faceting, core–shell formation, and amorphous-to-crystalline transitions. Comparative analyses across different synthesis techniques underscore the role of rapid quenching, nonequilibrium solidification, and high-entropy effects in dictating structural complexity. This chapter further explores *in situ* and combinatorial microscopy approaches that enable high-throughput mapping of microstructural libraries. We have divided the chapter into various sections focusing on characterization at different length scales, i.e., the nano-to-atomic scale. By providing insights into growth mechanisms and structure–property relationships, this work establishes microscopy as an indispensable tool for guiding the rational design of functional laser-synthesized nanoparticles in catalysis, energy, and quantum applications.

5.1 Nanoscale characterization

Laser-based synthesis has emerged as a versatile approach for producing nanoparticles with well-controlled size, morphology, and composition. Techniques such

doi:10.1088/978-0-7503-6224-5ch5

as pulsed LAL, pulsed laser deposition, and pulsed laser-induced dewetting (PLiD) offer unique advantages, including surfactant-free synthesis, rapid nonequilibrium processing, and access to metastable or multicomponent nanostructures [1–7]. A critical step in exploiting these materials for applications in catalysis, energy conversion, optics, or biomedical systems is their nanoscale characterization. Advanced microscopy and spectroscopy techniques enable researchers to unravel the structural, chemical, and functional properties that arise from laser–matter interactions at ultrafast timescales. While the nanoparticles formed by these techniques vary from a few nanometers to hundreds of nanometers, SEM is exhaustively utilized to study the nanoparticle formation ranging above 10 nm in size. In addition, the utilization of x-ray diffraction (XRD) provides average crystal structure and phase-related information in the nanoparticles over a large area.

Although there are several examples of SEM characterizations of nanoparticles, figure 5.1 illustrates the example of nanoparticle evolution via PLiD of Au thin film. As shown in the figure, SEM images demonstrate the progression of nanostructure morphologies as a function of the number of laser pulse irradiations that subsequently lead to the formation of well-separated droplet-shaped nanoparticles from a 12 nm thick parent thin film. The post-acquisition analysis of the obtained images

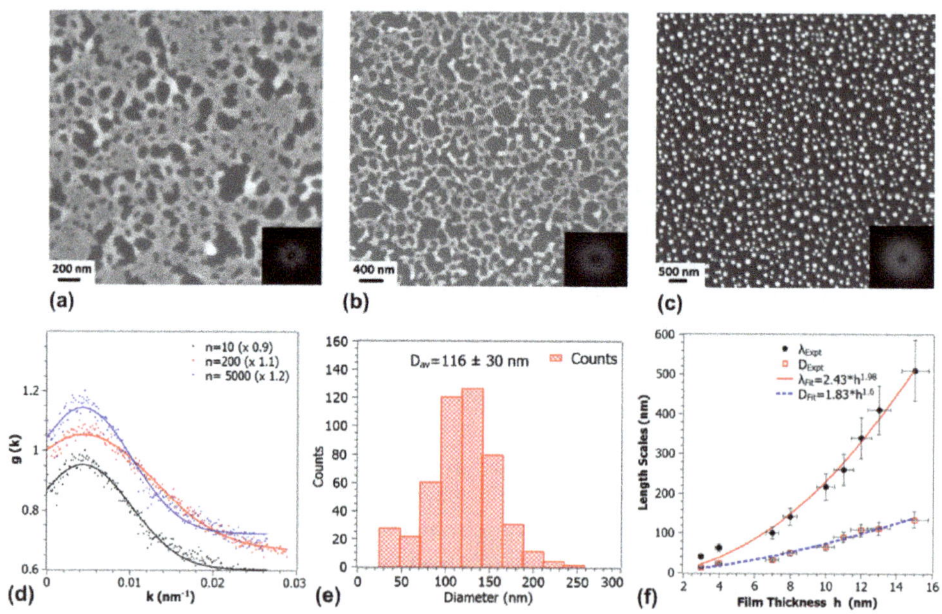

Figure 5.1. (a–c) SEM images of the progression of dewetting in 12 nm thick film with an increasing number of laser pulses *n* (10, 200, and 10 500). (d) Plot of the RDF for each stage of dewetting with Gaussian peak fitting. The peak position in the RDF was used to estimate the characteristic length scales. (e) Size distribution and average diameter of nanoparticles from the SEM image of (c). (f) Analysis of the various length scales in the final robust nanoparticle state as a function of film thickness. The fits show the exponent in good agreement with predictions from linear thin film hydrodynamic theory. Reproduced from [8], with permission from Springer Nature.

conducted by Yadavali *et al* provides the radial distribution function, which informs the average size and the nearest neighbor distance between the particles [8]. As the results show, the dewetting length scale (nearest neighbor distance) is determined to have a dependence of approximately h^2 on the parent film thickness (h), while the size dependence is approximately $h^{1.6}$. These dependences help in understanding the dewetting mechanisms (i.e., Raleigh Pleatue instabilities, the Marangoni effect, and the thermocapillary effect) that lead to the final distribution of nanoparticles via affecting the liquid state mass transport of metals [6, 9, 10].

Figure 5.2 demonstrates the examples of nanoparticle material libraries of various mono-metals, bimetals, and multi-principal element alloys (MPEA) formed via the PLiD of thin films. While there is substantial work performed in the monometallic (Ag, Au, Pt, Pd, Ni, Co, Fe, etc) and bimetallic systems, these are some of the

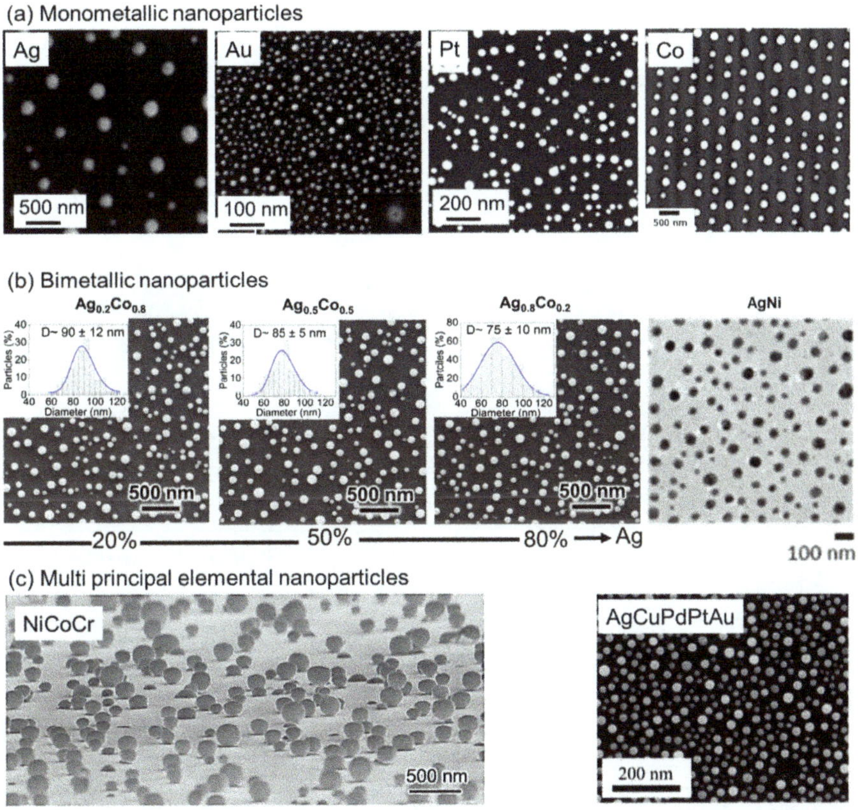

Figure 5.2. Representative SEM images of the nanoparticle material libraries of (a) monometallic (Ag, Au, Pt, and Co), [Reproduced from [11]. © IOP Publishing Ltd. All rights reserved; Reproduced from [8], with permission from Springer Nature; Reprinted with permission from [12]. Copyright (2025) American Chemical Society; Reprinted from [13], Copyright (2011), with permission from Elsevier.], (b) bimetallic ($Ag_{0.2}Co_{0.8}$, $Ag_{0.5}Co_{0.5}$, $Ag_{0.8}Co_{0.2}$, and $Ag_{0.5}Ni_{0.5}$) [[14–17] John Wiley & Sons. © 2024 Wiley-VCH GmbH], and (c) multi-principal elemental (NiCoCr and AgCuPdPtAu) nanoparticles formed by the PLiD method; left image reproduced from [16] John Wiley & Sons. © 2024 Wiley-VCH GmbH, right image reproduced from [17] CC BY 3.0.

representative examples presented in figure 5.2. These representative electron microscopy images highlight the morphological diversity and compositional tunability of laser-synthesized nanoparticles. Monometallic systems such as Ag, Au, Pt, and Co form well-dispersed hemispherical nanoparticles with sizes in the tens of nanometers [8, 11–13], while alloyed systems (e.g., Ag–Co, Ag–Ni, NiCoCr, and high-entropy AgCuPdPtAu) [14–17] demonstrate controlled variation in particle diameter, distribution, and ordering. The insets in Ag–Co nanoparticles, showing particle size histograms, illustrate the high uniformity achievable by laser-induced dewetting, as well as the compositional dependence of the average particle size. Collectively, these micrographs highlight the capability of laser synthesis to engineer nanoparticles, ranging from simple elemental systems to complex multi-component alloys, with precise nanoscale morphology.

Mandal and coworkers have also demonstrated distinguishing core–shell structures within the nanoparticles using SEM characterization as shown in figure 5.3 [18]. These results demonstrate the formation of core–shell nanoparticles with well-defined morphology and uniform coverage following laser-induced synthesis. SEM images (figure 5.3(a)) reveal spherical particles exhibiting a consistent contact angle of ~145°,

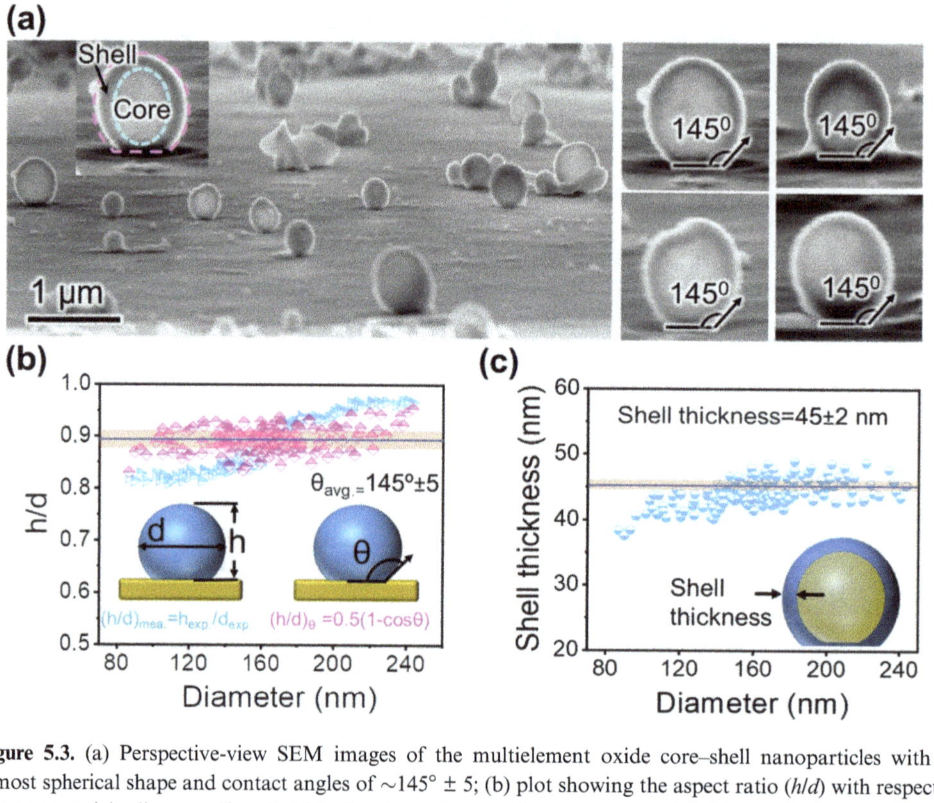

Figure 5.3. (a) Perspective-view SEM images of the multielement oxide core–shell nanoparticles with an almost spherical shape and contact angles of ~145° ± 5; (b) plot showing the aspect ratio (*h*/*d*) with respect to the nanoparticle diameter [*h* = height, *d* = diameter, and *h* = contact angle]; (c) plot presenting the shell thickness with respect to core–shell nanoparticle diameters. The average shell thickness is ~ 45 ± 2 nm. Reproduced from [18].

indicative of stable hemispherical geometry. Quantitative analysis (figure 5.3(b)) shows a nearly constant height-to-diameter ratio across a broad size distribution, confirming reproducible droplet-like shapes governed by surface energy minimization. Thickness measurements (figure 5.3(c)) further highlight the presence of a conformal shell with an average thickness of 45 ± 2 nm, independent of particle diameter. From the SEM images, it is observed that the nanoparticles exhibit two distinct contrasts in the inner and outer regions, where darker contrast consistently covers the outer region of each nanoparticle while the lighter contrast covers the core region, implying the formation of nanoparticles with a core–shell structure. Secondary electron (SE) imaging is used in this study, which is recognized to be more effective in obtaining contrast difference in core–shell nanoparticles. The topographic contrast of these core–shell nanoparticles is more prominent in SE images, as the SE escape depth is shallower than that of backscattered electrons in such conditions. In principle, the observed contrast in these core–shell nanoparticles can be explained based on the atomic mass and density of the elements.

In addition to SEM characterization, XRD has been extensively used to determine the crystal structure and related strain in the nanoparticles at the bulk level. XRD analysis (figure 5.4(a)) by [19] reveals the structural evolution of bimetallic nanoparticles on ZnO supporting substrate. Characteristic reflections confirm the ZnO substrate with metallic Au and Pd phases, while systematic peak shifts in the (111) reflection indicate lattice distortion arising from alloying between Au and Pd. Quantitative analysis (figure 5.4(b)) shows a linear correlation between the Pd molar fraction and the interplanar spacing of (111) planes of the metals, consistent with Vegard's law, thereby confirming the formation of Au–Pd solid solutions within the ZnO matrix. These results highlight the ability of laser processing to drive compositional tuning and controlled alloying in multicomponent oxide–metal nanostructures.

5.2 High-resolution nanoscale and atomic-level structural and chemical characterization

High-resolution characterization techniques provide essential insights into the nanoscale morphology, crystallography, and chemistry of metal nanoparticles synthesized by laser-based methods. The rapid quenching and nonequilibrium nature of pulsed laser processing often produce unique structural features such as lattice distortions, defect-rich domains, metastable phases, and compositional gradients that cannot be captured by conventional bulk analysis.

Nanoparticle imaging has been commonly done using scanning/transmission electron microscopy to investigate nanoparticle structure and chemical information. Several groups have reported investigations ranging from simple nanoparticle morphologies, statistical particle size analysis, to more intriguing phase transitions, surface valence state variation, and chemical distribution. For example, Zhang and coworkers have reported TEM studies on the formation of MnO_2 nanoparticles using the LAL process, which leads to phase transformation from Mn_3O_4 to Mn_2O_3

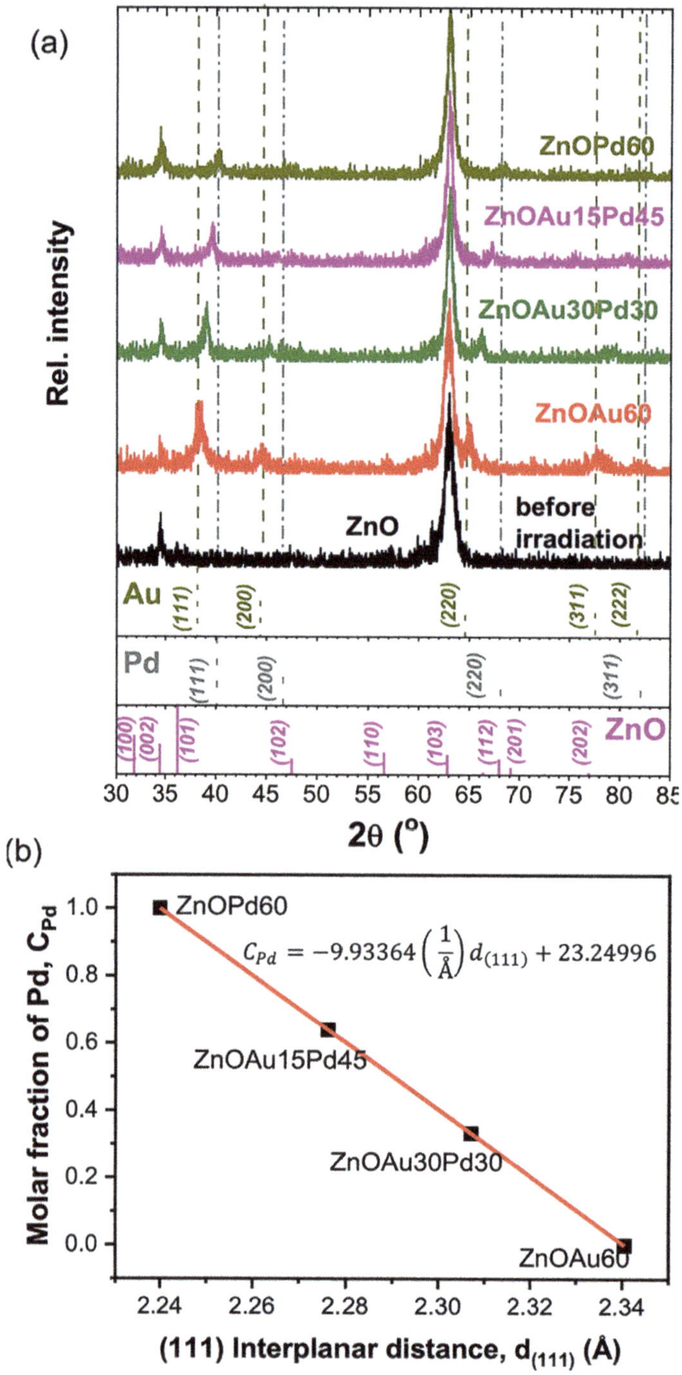

Figure 5.4. (a) XRD patterns of a ZnO film before irradiation and ZnO films with Au, Pd, and Au–Pd nanoparticles, (b) Pd molar fraction CPd of the samples after irradiation, which was calculated applying Bragg's and Vegard laws using the 2θ angle positions corresponding to the (111) planes of metals. Reprinted from [19], Copyright (2023), with permission from Elsevier.

and then to γ–MnO_2 as the ratio of ethanol/water is changed in the liquid media during ablation [20].

Via atomic-resolution investigations, Xi–Wen Du and coworkers reported a grain-boundary driven enhancement in the oxygen evolution reaction performance in LAL-synthesized Ru nanoparticles [21]. Figure 5.5 highlights high-resolution structural and functional characterization of laser-synthesized Ru (L-Ru) nanoparticles. Low-magnification TEM (figure 5.5(a)) reveals a uniform particle size distribution centered around 10–12 nm, while atomic-resolution imaging (figure 5.5 (b, c)) resolves distinct grain boundaries, triple junctions, and lattice fringes corresponding to the Ru(101) plane with an interplanar spacing of 0.206 nm. These nanoscale features directly influence catalytic behavior, as shown by electrochemical measurements (d), where L-Ru exhibits markedly higher specific activity compared to annealed Ru and commercial RuO_2. Electrochemical impedance spectra (e) confirm reduced charge-transfer resistance for L-Ru, underscoring its superior conductivity and reaction kinetics. Optical absorption spectra (f) further demonstrate the distinct electronic structure of L-Ru, with enhanced intensity relative to other benchmarks. Collectively, these results establish a strong link between atomic-level structure and catalytic performance, underscoring the advantages of laser processing for designing high-activity Ru nanocatalysts.

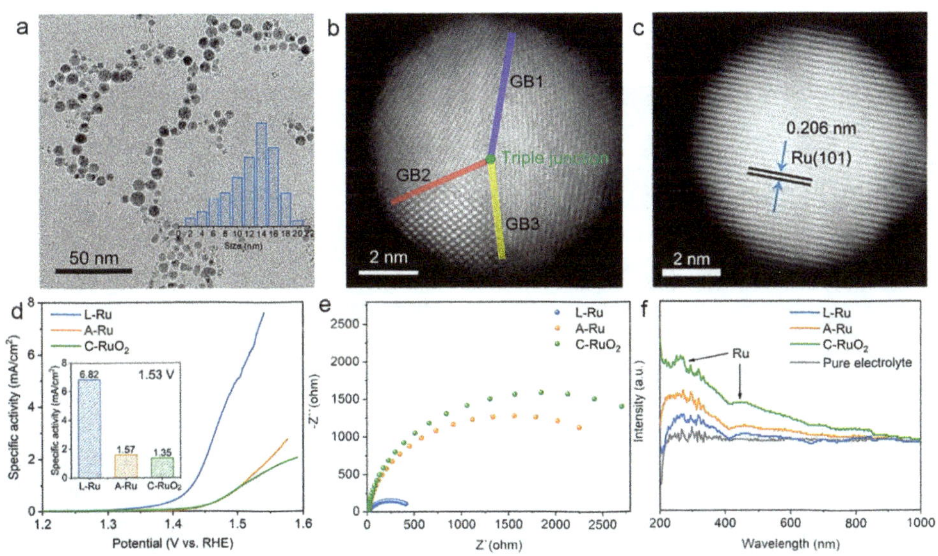

Figure 5.5. Grain-boundary-rich metal particles and their OER performance. (a) TEM image of Ru particles synthesized by the LAL of Ru in water (the inset image is the particle histogram). (b) and (c) high-angle annular dark-field scanning transmission electron microscopy (HAADF-STEM) images of laser-synthesized Ru (L-Ru) particles with three-grain boundaries and a Ru particle annealed (A-Ru) by heat-treating L-Ru at 673 K for 3 h in N2. (d–f) OER performance of electrodes using L-Ru, A-Ru, and commercial RuO_2 (C-RuO_2). (d) Linear-sweep voltammetry plots corrected by electrochemical surface area; the inset shows specific activity at 1.53 V versus reversible hydrogen electrode (RHE). (e) Electrochemical impedance spectroscopy curves at 1.387 V. (f) UV–vis of electrolytes after 10 h chronopotentiometry tests with L-Ru, A-Ru, and RuO_2. Reprinted with permission from [21]. Copyright (2020) American Chemical Society.

In addition to atomic-resolution imaging, quantitative chemical analyses have been extensively studied using TEM, which provides insights into elemental homogeneity/inhomogeneity, spatial distribution, and composition. For instance, energy-dispersive x-ray spectroscopy (EDS) in TEM enables mapping of elemental distributions across individual nanoparticles, allowing researchers to identify compositional uniformity, alloying behavior, or the presence of core–shell structures. One of the examples is presented in figure 5.6, illustrating high-angle annular dark-field (HAADF) images and EDS-based elemental analysis of Au–Fe nanoparticles produced using LAL in different liquid media. As discussed in this work by Coviello *et al,* LAL of Au–Fe systems produces a rich variety of heterostructures, including crescents, core–shells, and Janus nanoparticles, whose composition and interfaces are strongly dependent on the surrounding liquid environment [22]. HAADF images revealed distinct z-dependent intensity contrast corresponding to Fe_3O_4 and FCC Au, confirming the coexistence of iron oxide and metallic gold regions within the same nanoparticles. EDS elemental mapping and line scans further substantiated these findings, providing nanoscale chemical resolution of Au, Fe, and O distributions. These results point to dealloying and oxidation processes of Fe atoms initially present in an Au–Fe nanophase formed during the earliest stage of laser irradiation, when atomic mobility is still high, leading to sharp metal/oxide interfaces.

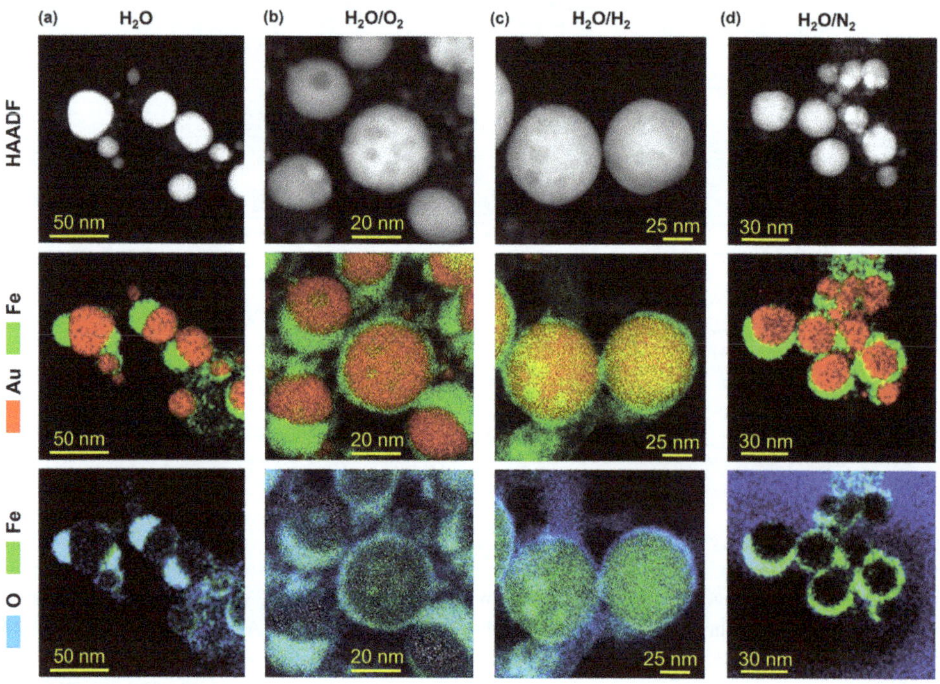

Figure 5.6. STEM-HAADF (top), EDX map of Au M and Fe K lines (middle), and EDX map of Fe and O K lines (bottom) of representative nanoparticles from the H_2O a), H_2O/O_2 b), H_2O/H_2 c), and H_2O/N_2 d) samples. Reproduced from [22]. CC BY 4.0.

Comparative analysis of samples synthesized in different liquid environments highlights the influence of solution chemistry on nanoparticle structure and composition. In the H_2O/O_2 environment, incomplete dealloying resulted in a hybrid ultrastructure composed of oxidized Fe domains embedded within a bimetallic Au–Fe matrix, surrounded by oxide crescents and shells. In contrast, the H_2O/H_2 samples exhibited a thin iron oxide shell with Fe-rich clusters within the Au–Fe matrix, with a significantly higher than average Fe content (~36 at.%) detected by EDX, exceeding values inferred from XRD. This discrepancy, together with localized oxygen signals, suggests the precipitation of metallic Fe alongside iron oxides in the H_2O/H_2 environment. Overall, these results demonstrate how liquid environments dictate the extent of Fe dealloying and oxidation, producing nanoscale heterostructures with unique phase coexistence and compositional heterogeneity.

Elemental analysis, critical to nanomaterials design, has also been used in bi- and multimetallic nanoparticles formed by PLiD of the films. Figure 5.7 presents the TEM/STEM–EDS analysis of laser-dewetted Au–Sn nanoparticles reported by Dzienny *et al,* highlighting the structural and compositional diversity that arises from differences in thin film preparation [23]. For the Si/Sn (11.4 nm)/Au (3.6 nm) bilayer sample (figure 5.7(a)), the elemental mapping shows a relatively uniform distribution of Au and Sn within the nanoparticle, with an estimated stoichiometric ratio of ~1:1, consistent with the formation of Au–Sn intermetallic phases. In contrast, the Si/Au (0.6 nm)/Sn (11.4 nm)/Au (3.6 nm) thin film sample (figure 5.7(b)) reveals pronounced compositional inhomogeneity, with Sn-rich regions and localized oxygen incorporation suggestive of SnO_2 formation at the particle–substrate interface. The inset in figure 5.7(b) further illustrates nanoscale element maps, clearly delineating Au- and Sn-rich domains and confirming the multiphase character of these particles. Together, these results demonstrate that variations in deposition sequence and underlying layer structure critically influence the alloying, oxidation, and nanoscale phase distribution within laser-synthesized nanoparticles.

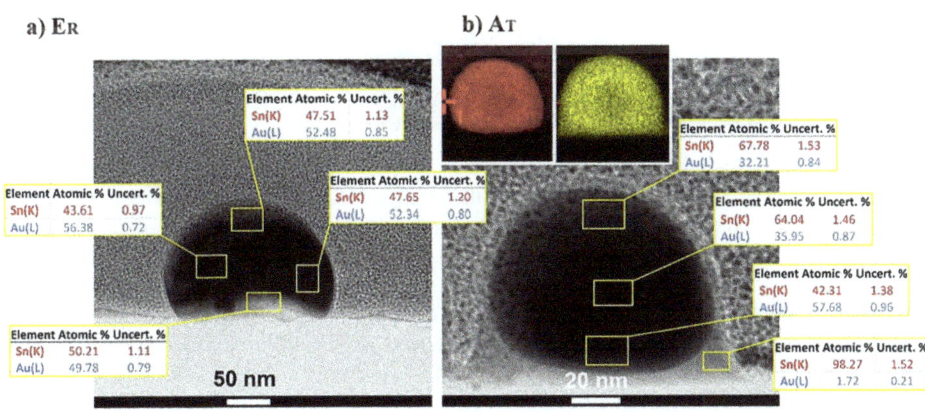

Figure 5.7. STEM-EDS analysis of samples E_R (a) and AT (b) with estimated Au:Sn ratios for selected areas of nanoparticles. Insert in *b* is the nanoscale element mappings of a single A_T particle. Reproduced from [23].

In addition to EDS, electron energy-loss spectroscopy (EELS) has been used to chemically characterize the nanoparticles. While EDS and EELS methods are both capable of producing elemental distribution, EELS also provides quantitative information on the valence state of constituent elements, their bond characteristics, local atomic environment, and electronic structure. Sachan and coworkers have demonstrated the elemental distribution of Ag and Co in bimetallic nanoparticles synthesized by the PLiD process. Figure 5.8 illustrates the nanoscale structural and compositional characteristics of Ag–Co nanoparticles synthesized through bilayer self-organization of Co/Ag thin films [24]. The SEM image (a) confirms the uniform arrangement of nanoparticles with a narrow size distribution around 125 ± 20 nm, and the fast Fourier transform inset highlights short-range spatial ordering across the array. A cross-sectional HAADF-STEM micrograph (b) reveals nearly hemispherical nanoparticles with internal contrast variation, suggesting polycrystallinity and phase separation within individual particles. EELS spectra recorded at selected

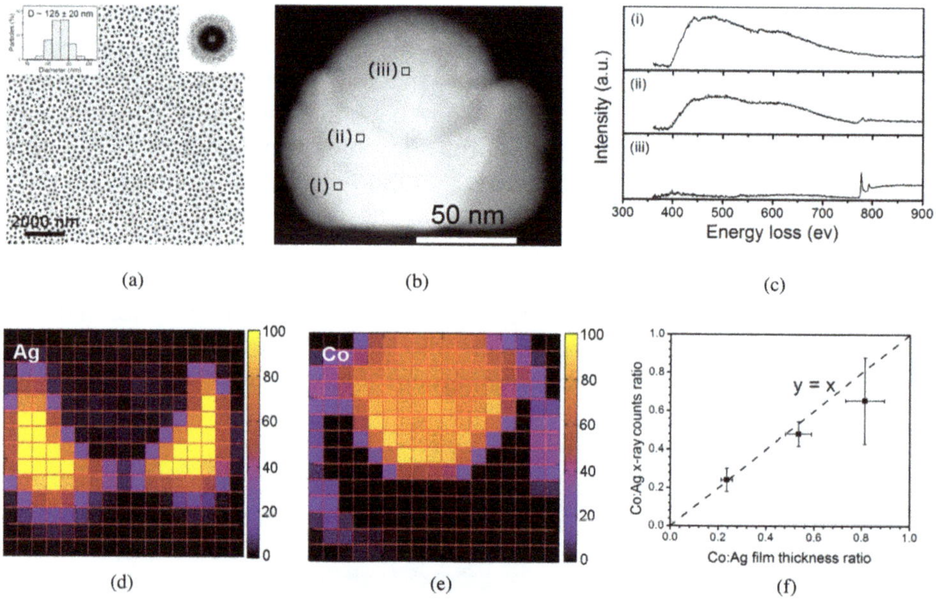

Figure 5.8. (a) A typical SEM micrograph showing the arrangement of nanoparticles following bilayer self-organization from the Co (5 nm)/Ag (5 nm)/SiO_2 bilayer. The histogram in the inset shows the narrow size distribution of particles achieved by this process. The fast Fourier transform image in the inset shows the spatial short-range ordering between the nanoparticles. (b) A cross-sectional HAADF image of a nearly hemispherical-shaped Ag–Co nanoparticle made from the Co (5 nm)/Ag (5 nm)/SiO_2 bilayer showing contrast variation indicating polycrystallinity within the particle. (c) Background-subtracted EELS spectra at different regions of the nanoparticle in (b), showing delayed Ag M4,5 edge energy at 394 eV and Co L3 edge energy at 779 eV. (d) Ag and (e) Co EELS compositional maps of the enclosed region shown in image (b), exhibiting the immiscibility of Co and Ag in each other. The step size of the compositional map is 5.9 nm × 5.9 nm. The contrast bar shows the variation of atom % of Ag and Co individually in different locations of the enclosed region of image (b). (f) The average Co:Ag ratio in each nanoparticle from x-ray mapping (symbols) plotted against the Co:Ag film thickness ratio. Reproduced from [24].

regions (c) identify the Ag $M_{4,5}$ edge and the Co L_3 edge, thereby confirming the coexistence of both metals within the same nanoparticle but with spatially distinct domains. This immiscibility is further visualized in EELS elemental maps (d, e), which show complementary spatial distributions of Ag- and Co-rich regions at the nanoscale. Quantitative analysis (f) demonstrates good correlation between the nominal Co:Ag film thickness ratio and the average Co:Ag composition ratio determined by x-ray mapping, validating the control of nanoparticle stoichiometry through bilayer thickness engineering. Together, these results provide compelling evidence of compositionally phase-separated but structurally integrated Ag–Co nanoparticles formed by laser dewetting, emphasizing the role of immiscibility in dictating final nanoscale architectures.

Another example of high-resolution elemental analysis is presented by Mandal and coworkers, who demonstrated the use of atom probe tomography (APT) to characterize a single nanoparticle and understand the elemental and compositional distribution [18]. Figure 5.9 illustrates the atomic-scale characterization of a core–shell nanoparticle using APT. A focused ion beam lift-out (a) shows the targeted

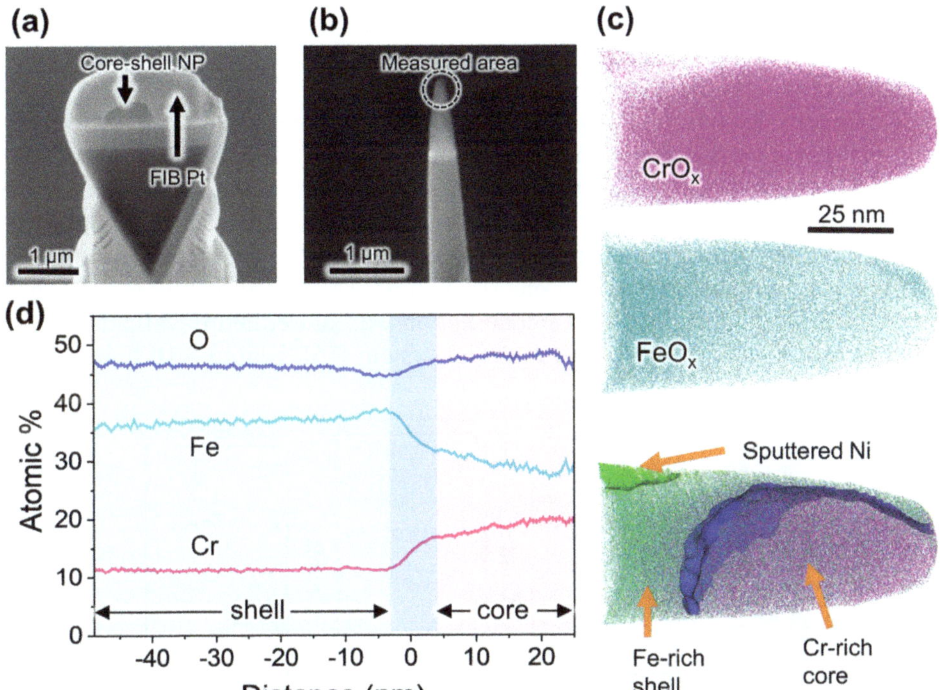

Figure 5.9. Cross-sectional SEM image of (a) the wedge liftout showing the targeted core–shell nanoparticle covered with focused ion beam (FIB) Pt; (b) final APT specimen with the core–shell nanoparticle at needle apex. The dotted circle represents the approximate volume of the APT measurement; (c) APT atom maps showing CrO_x and FeO_x ions separately and together with Ni ions and 20 at.% Cr and Ni isoconcentration surfaces; (d) proximity histogram presenting the at.% of Fe, Cr, and O in both the core and shell region of a core–shell nanoparticle with a diffused interfacial region. Reproduced from [18]. CC BY 4.0.

nanoparticle protected by sputtered Ni and Pt layers, while the prepared sharp needle specimen (b) exposes the nanoparticle at the tip for APT analysis. The atom maps in (c) clearly reveal a core–shell configuration, with Cr-oxide concentrated in the core and Fe-oxide enriched in the shell, while solute elements such as Ni, Co, and Mn remain uniformly distributed throughout. The proximity histogram across the 20 at.% Cr iso-concentration surface (d) quantifies this segregation, showing Fe enrichment (~35 at.%) in the shell and Cr enrichment (~20 at.%) in the core, with oxygen content (~45 at.%) nearly constant in both regions. These findings indicate preferential partitioning of Cr into the nanoparticle core and Fe into the shell during ultrafast laser-induced melting and resolidification. The rapid solidification dynamics (within ~100 ns per laser pulse) drive elemental redistribution and phase segregation, ultimately yielding stable core–shell nanoparticles with distinct Cr-oxide-rich cores and Fe-oxide-rich shells.

In addition to elemental characterization, the EELS technique is also used for probing surface plasmons in the optically active nanoparticles. Various examples of localized surface plasmon resonance have been shown in Ag–Co, Ag–Ni and Ag–Au nanoparticles, where EELS methodology in the low-loss range (0–10 eV) provides critical information with high energy and spatial resolution. Figure 5.10 illustrates such an example of size-dependent localized surface plasmon resonance (LSPR) evolution in PLiD-synthesized Ag–CoFe nanoparticle, which makes an elementally segregated Janus-like structure at the nanoscale [25].

This work by Sachan and coworkers, for the first time, reports the evolution of ferroplasmons in bimetallic nanoparticles [25]. These plasmons are observed in ferromagnetic nanostructures in the visible spectrum with comparable intensity and bandwidth to those of the LSPRs from the Ag regions and are termed ferroplasmons for short [25–27]. Usually, ferromagnetic materials (Co, Ni, Fe, etc) show poor or no plasmonic properties, unlike noble metals such as Ag, Au, and Pt that have highly intense LSPR characteristics. This phenomenon was subsequently validated by P D Rack and coworkers in the Ag–Ni material system [15]. Figure 5.10 highlights the

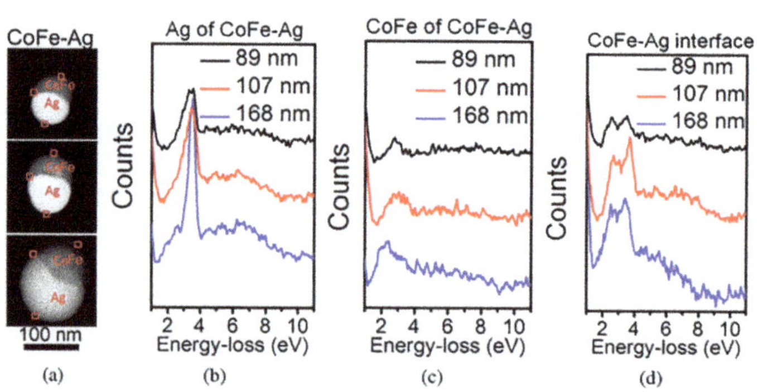

Figure 5.10. (a) HAADF image of various CoFe–Ag nanoparticles on the C-substrate, where the size of the nanoparticle increases from top to bottom. The scale bar of 100 nm is the same for all images. EELS spectra from (b) the Ag side, (c) the CoFe side, and (d) the interface between the CoFe and Ag of the CoFeAg nanoparticles, as a function of particle diameter. Reprinted with permission from [25]. Copyright (2014) American Chemical Society.

nanoscale analysis of CoFe–Ag Janus or dumbbell-shaped nanoparticles, revealing distinct spatial and spectral plasmonic features. STEM imaging (a) shows the distribution of Ag and CoFe regions within the nanoparticles, enabling site-specific EELS investigations.

The EELS spectra figure 5.10(b–d) capture LSPRs at the Ag-rich domain, the CoFe-rich domain, and at the CoFe–Ag interface for nanoparticles of different sizes (89, 107, and 168 nm). In the Ag domains (b), strong plasmon peaks are observed at ~3–4 eV, characteristic of metallic silver. The CoFe regions (c) exhibit damped and broadened plasmonic features, a consequence of higher intrinsic losses from ferromagnetic d-electrons. At the CoFe–Ag interface (d), hybridized spectral features emerge, reflecting plasmon–ferromagnet coupling—this mixed response underpins the ferroplasmonic behavior. Importantly, the spectra shift broadens as the particle size increases, confirming that both geometry and interface quality strongly influence plasmonic resonance.

These results demonstrate that embedding ferromagnetic CoFe within a plasmonic Ag matrix produces hybridized excitations, ferroplasmon, which combine optical tunability with magnetic responsiveness. Such ferroplasmonic nanoparticles open pathways for multifunctional applications, including magneto-plasmonic sensing, photothermal therapy with magnetic guidance, and data storage technologies where optical and magnetic modalities are coupled.

5.3 *In situ* characterizations

Researchers have utilized *in situ* TEM techniques to understand the evolution of nanoparticles, such as in the PLiD method. One example is reported by McKeown and coworkers, who demonstrated the formation of Cu–Ni bimetallic nanoparticles using time-resolved imaging, as shown in figure 5.11 [28]. The figure illustrates *in situ*, time-resolved imaging of laser-induced dewetting dynamics in a 10 nm thick Co–Cu bilayer thin film on a silicon nitride substrate, highlighting two distinct dewetting pathways. Figures 5.11(a) and (b) present sequential images captured at the center of a Gaussian laser heating pulse, with deposited energies of 4.3 and 5.2 μJ, respectively. At lower energy (a), dewetting proceeds gradually, with morphological instabilities forming at ~50–100 ns, followed by the nucleation and growth of isolated nanoparticles over several hundred nanoseconds. In contrast, at higher energy (b), the dewetting process is markedly accelerated, with nanoparticle formation observable within tens of nanoseconds, leading to a denser and more uniform nanoparticle array.

The energy-filtered TEM images and elemental maps of the final nanoparticle assemblies in lower panel confirm the formation of Co–Cu core–shell structures. Zero-loss images reveal uniform arrays of spherical particles, while overlaid elemental maps show a distinct distribution of Co (cyan) in the cores and Cu (red) in the shells. This spatial segregation reflects differential diffusion and surface energetics during ultrafast melting and resolidification. Collectively, the results demonstrate how laser pulse energy not only controls the kinetics of thin-film dewetting but also governs nanoscale elemental redistribution, thereby enabling

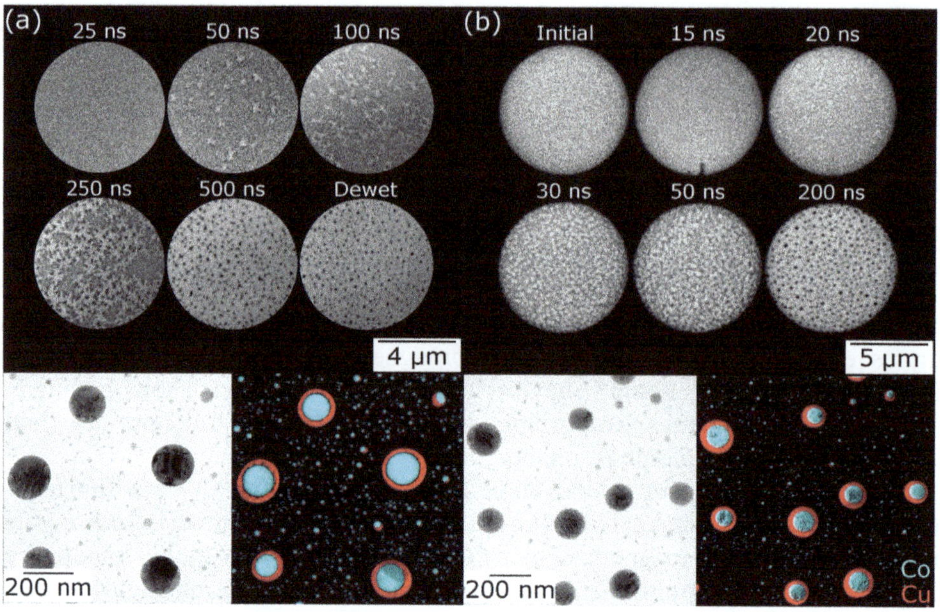

Figure 5.11. *In situ* time-resolved imaging of dewetting via two mechanisms, with associated elemental maps of resultant core–shell particle arrays. Time-delay series of images as a 10 nm thick Co–Cu film dewets a silicon nitride substrate. The image series in (a) and (b) were acquired at the center of the Gaussian laser heating pulse with total deposited energies of (a) 4.3 μJ and (b) 5.2 μJ. Beneath each of the time-delay series are energy-filtered TEM images of the resultant core–shell particles: zero-loss image and associated overlaid Co and Cu maps. [28] John Wiley & Sons. © 2014 WILEY-VCH Verlag GmbH & Co. KGaA, Weinheim

tailored synthesis of core–shell particle arrays with compositional and structural precision.

Klaus van Benthem and coworkers also show the dewetting of Au–Ni bilayer film by cross-sectional bright-field TEM micrographs (figure 5.12) in their as-deposited (a, b) and annealed (c, d) states, highlighting the nanoscale morphological evolution at the interface [29]. In the as-deposited condition, Au islands are observed resting on the Ni substrate, with a distinct void forming at the triple junction between the Au grain boundary and the Ni surface (b). This interfacial instability reflects the competition between surface and grain boundary energies, which drive localized void formation and grooving at the metal–substrate junction.

Upon annealing, significant morphological changes are observed in both the Au and Ni layers (c, d). The Au islands undergo faceting, as indicated by the arrows, consistent with the minimization of surface energy under thermal treatment. Simultaneously, the Ni layer exhibits pronounced interfacial grooving beneath the Au islands, suggesting enhanced diffusional mass transport along the grain boundary and surface. The combination of faceting in the Au and grooving in the Ni highlights the dynamic interplay of capillarity-driven processes and diffusion kinetics during annealing. The presence of an amorphous carbon overlayer, originating from contamination during TEM preparation, does not obscure the key interfacial features but is noted for completeness.

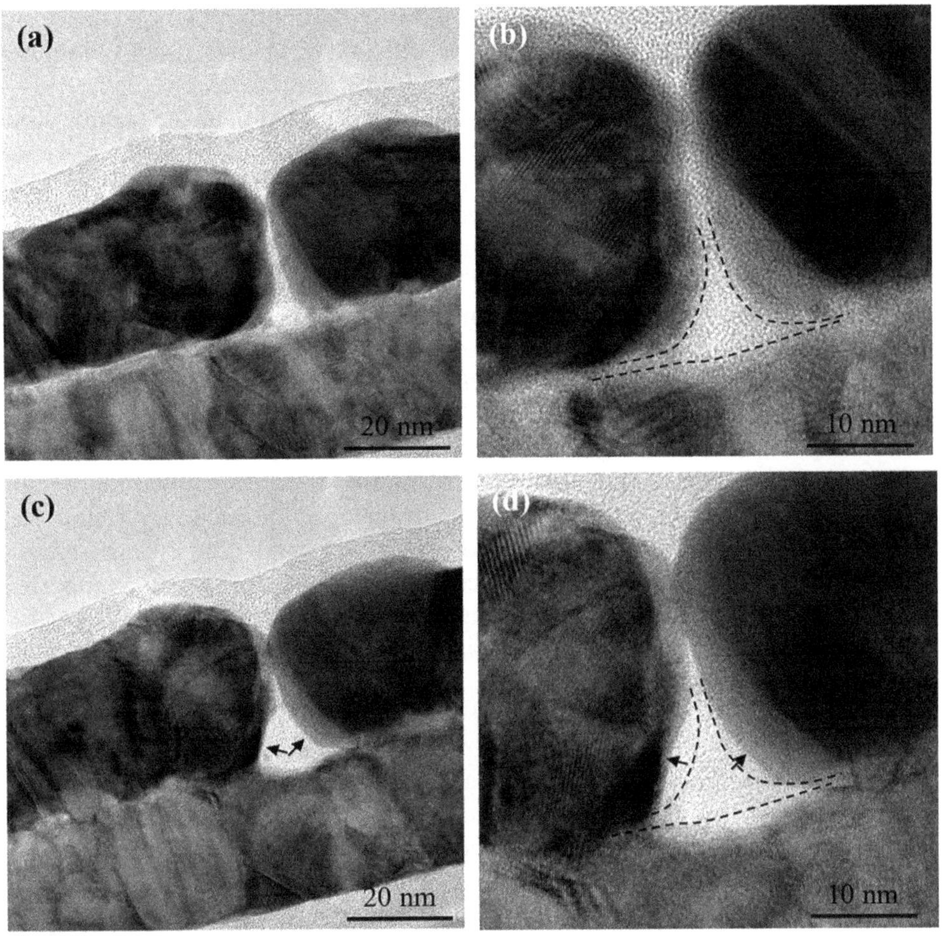

Figure 5.12. Cross-sectional bright field TEM micrographs of as-deposited (a, b) and the annealed (c, d) bilayer film. The figures show the same two Au islands with a void opening up at the triple line between a grain boundary in the Au layer and the Ni layer surface. After annealing, both Au islands reveal the formation of facets (arrows), while grooving occurs within the underlying surface of the Ni layer. The bilayer film was covered with amorphous carbon due to some contamination in the TEM. Reprinted from [29], Copyright (2017), with permission from Elsevier.

5.4 Data-science-driven characterization

The use of data science to quantitatively investigate specifically the laser-synthesized nanoparticles is critical, yet still very limited. Researchers have explored data-driven four-dimensional scanning transmission electron microscopy (4D-STEM) to study the crystal structure, phase formation, and lattice strain in the nanoparticles [16, 30, 31]. The advent of 4D-STEM has revolutionized nanoscale characterization by enabling simultaneous acquisition of real-space and reciprocal-space information across extended fields of view. This technique is particularly powerful for studying metal nanoparticles synthesized via laser-based methods, where rapid melting, quenching, and dewetting processes often yield nonequilibrium structures, complex phase mixtures, and subtle lattice distortions that are difficult to resolve using conventional imaging or diffraction.

In 4D-STEM, a finely focused electron probe is scanned over the sample while a pixelated detector records a full diffraction pattern at each probe position, generating a rich four-dimensional dataset (two real-space + two reciprocal-space dimensions). When applied to laser-synthesized nanoparticles, this multidimensional approach provides unprecedented insight into nanoscale structural heterogeneity [32, 33]. Data-driven analysis methods, including principal component analysis, non-negative matrix factorization, and machine-learning–based clustering, can be employed to extract hidden patterns from large datasets [32]. Such approaches reveal variations in local crystallographic orientation, strain fields, and defect distributions across individual nanoparticles and particle ensembles.

For example, phase mapping of bimetallic or high-entropy nanoparticles produced by laser ablation or pulsed laser dewetting can uncover compositional inhomogeneities and nanoscale segregation not visible in averaged diffraction profiles. Strain mapping derived from 4D-STEM further identifies local distortions arising from rapid quenching or lattice mismatch at interfaces in core–shell and Janus nanoparticles. Moreover, analysis of diffuse scattering provides information on short-range order and disorder, offering valuable insight into the metastable phases stabilized by ultrafast laser synthesis.

The synergy between 4D-STEM and advanced computational methods thus enables a data-driven framework for quantitative nanoscale characterization. Beyond simple visualization, these approaches allow researchers to correlate nanoparticle synthesis conditions with emergent structural motifs, to build predictive models of microstructural evolution, and to uncover statistical distributions of crystallography and strain over thousands of particles. In this way, 4D-STEM represents a bridge between traditional electron microscopy and modern data science, positioning itself as a key tool for unraveling the complexity of laser-synthesized metal nanoparticles.

In laser-synthesized Ag–Au nanoparticles, strain mapping has revealed pronounced lattice distortions at the Ag–Au interfaces, where differences in atomic size and lattice parameter ($\sim$4% mismatch) lead to coherent or semi-coherent interfaces. These interfacial strains are not uniformly distributed but are concentrated at facets, edges, and twin boundaries, where strain relaxation mechanisms such as dislocations or stacking faults can be observed. The ability to visualize these strain fields provides a direct link to the plasmonic behavior of Ag–Au nanoparticles, as interfacial strain modifies conduction electron density and thus shifts LSPRs. Data-driven 4D-STEM analyses further highlight heterogeneity across particle ensembles, showing that smaller nanoparticles tend to sustain higher elastic strains due to limited dislocation-mediated relaxation pathways.

For high-entropy nanoparticles synthesized by laser ablation or pulsed laser dewetting, strain mapping uncovers even more complex behavior. The random mixing of multiple elements (e.g., Ag, Au, Pt, Pd, Cu) leads to significant local lattice distortions, often exceeding several percent, arising from variations in atomic radii and bonding environments. 4D-STEM strain maps show patchy strain fields with nanoscale fluctuations that cannot be captured by conventional XRD or averaged diffraction. Importantly, these heterogeneous strain distributions correlate with enhanced catalytic activity, as strain fields modulate surface electronic states and create defect-rich, high-energy sites. The ability of laser synthesis to 'freeze in' such nonequilibrium strain states during rapid quenching distinguishes it from equilibrium synthesis methods, which typically allow relaxation to lower-energy configurations.

Collectively, strain mapping of laser-synthesized Ag–Au and high-entropy nanoparticles provides a quantitative window into how ultrafast synthesis drives structural complexity, revealing interfacial coherency, nanoscale heterogeneity, and atomic-level lattice distortions. These findings underscore the role of local strain as a tunable parameter linking synthesis conditions to emergent optical, electronic, and catalytic properties, and highlight the importance of data-driven 4D-STEM in correlating processing, structure, and function at the nanoscale.

The latest use of 4D-STEM is demonstrated by Sachan and colleagues in NiCoCr MPEA nanoparticles, which is the parent domain of high entropy alloys. Figure 5.13 demonstrates the application of 4D-STEM for nanoscale crystallographic analysis of laser-synthesized nanoparticles [16]. In this approach (panel A), a focused electron probe is raster-scanned across the particle, with a nanodiffraction pattern recorded at each pixel, generating a rich four-dimensional dataset that captures both spatial and reciprocal-space information. By correlating these patterns, pixel-by-pixel diffraction intensities are extracted, allowing quantitative mapping of different crystallographic orientations.

Panel B highlights representative nanodiffraction patterns obtained from distinct grains within a single nanoparticle. By assigning intensity distributions from selected diffraction spots to different color channels, individual grains are

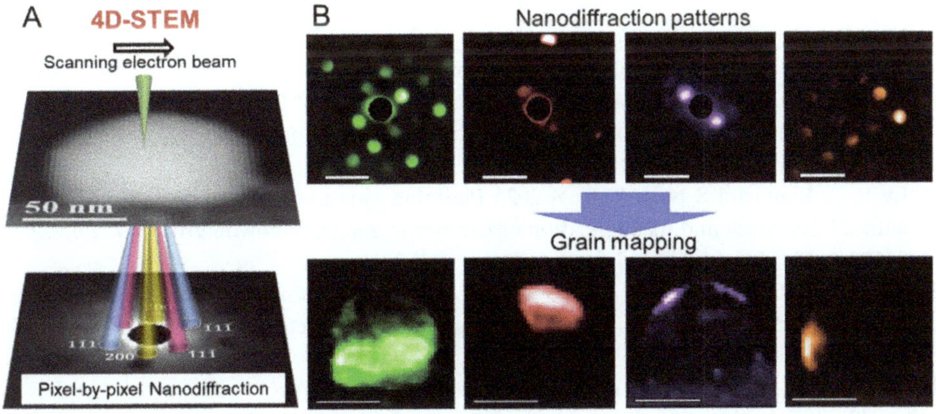

Figure 5.13. (A) Schematic of 4D-STEM showing cross-sectional image of NiCoCr nanoparticle and cumulative nanodiffraction pattern, (B) obtained unique nanodiffraction patterns and corresponding crystal grain mapping in the nanoparticle. [16] John Wiley & Sons. © 2024 Wiley-VCH GmbH

mapped with high spatial fidelity. The resulting orientation maps clearly reveal the presence of multiple crystalline domains and grain boundaries within a single nanoparticle, emphasizing the polycrystalline nature often associated with rapid, nonequilibrium laser-synthesis processes. This grain-mapping capability of 4D-STEM provides a powerful tool to visualize nanoscale heterogeneity, enabling correlations between synthesis conditions, crystallography, and functional performance.

Bibliography

[1] Forsythe R C, Cox C P, Wilsey M K and Mu A M 2021 Pulsed laser in liquids made nanomaterials for catalysis *Chem. Rev.* **121** 7568–637

[2] Zeng H, Du X-W, Singh S C, Kulinich S A, Yang S, He J and Cai W 2012 Nanomaterials via laser ablation/irradiation in liquid: a review *Adv. Funct. Mater.* **22** 1333–53

[3] Zhang D, Li Z and Sugioka K 2021 Laser ablation in liquids for nanomaterial synthesis: diversities of targets and liquids *J. Phys. Photonics* **3** 042002

[4] Thiele U, Velarde M G and Neuffer K 2001 Dewetting: film rupture by nucleation in the spinodal regime *Phys. Rev. Lett.* **87** 016104/1–4

[5] Gentili D, Foschi G, Valle F, Cavallini M and Biscarini F 2012 Applications of dewetting in micro and nanotechnology *Chem. Soc. Rev.* **41** 4430–43

[6] Kondic L, Gonzalez A G, Diez J A, Fowlkes J D and Rack P 2020 Liquid-state dewetting of pulsed-laser-heated nanoscale metal films and other geometries *Annu. Rev. Fluid Mech.* **52** 235–62

[7] Sachan R, Ramos V, Malasi A, Yadavali S, Bartley B, Garcia H, Duscher G and Kalyanaraman R 2013 Oxidation-resistant silver nanostructures for ultrastable plasmonic applications *Adv. Mater.* **25** 2045–50

[8] Yadavali S, Khenner M and Kalyanaraman R 2013 Pulsed laser dewetting of Au films: experiments and modeling of nanoscale behavior *J. Mater. Res.* **28** 1715–23

[9] Shirato N, Krishna H and Kalyanaraman R 2010 Thermodynamic model for the dewetting instability in ultrathin films *J. Appl. Phys.* **108** 024313

[10] Krishna H, Shirato N, Favazza C and Kalyanaraman R 2009 Energy driven self-organization in nanoscale metallic liquid films *Phys. Chem. Chem. Phys.* **11** 8136–43

[11] Krishna H, Sachan R, Strader J, Favazza C, Khenner M and Kalyanaraman R 2010 Thickness-dependent spontaneous dewetting morphology of ultrathin Ag films *Nanotechnology* **21** 155601

[12] Ly L Q, Bonvicini S N and Shi Y 2025 Platinum nanoparticle formation by pulsed laser-induced dewetting and its application as catalyst in silicon nanowire growth *J. Phys. Chem. C* **129** 4553–64

[13] Krishna H, Gangopadhyay A K, Strader J and Kalyanaraman R 2011 Nanosecond laser-induced synthesis of nanoparticles with tailorable magneticanisotropy *J. Magn. Magn. Mater.* **323** 356–62

[14] Paduri V R, Harimkar S and Sachan R 2024 Bimetallic AgCo nanoparticle synthesis via combinatorial nanosecond laser-induced dewetting of thin films *Adv. Eng. Mater.* **26** 2401258

[15] Garfinkel D A, Pakeltis G, Tang N, Ivanov I N, Fowlkes J D, Gilbert D A and Rack P D 2020 Optical and magnetic properties of Ag-Ni bimetallic nanoparticles assembled via pulsed laser-induced dewetting *ACS Omega* **5** 19285–92

[16] Mandal S, Gupta A K, Konečná A, Shirato A, Hachtel J A and Sachan R 2024 Creation of multi-principal element alloy NiCoCr nanostructures via nanosecond laser-induced dewetting *Small* **20** 2309574

[17] Ghebretnsae S, Ziehl T J, Purdy S, Zhang P, Sham T K and Shi Y 2025 Pulsed laser-induced dewetting for the production of noble-metal high-entropy-alloy nanoparticles *Nanoscale* **17** 15423–35

[18] Mandal S, Gupta A K, Echeverria E, McIlroy D N, Poplawsky J D and Sachan R 2022 Laser-assisted nanofabrication of multielement complex oxide core-shell nanoparticles *Mater. Des.* **220** 110882

[19] Segura-Zavala J, Depablos-Rivera O, García-Fernández T, Bizarro M, García-Morales R E and Sánchez-Aké C 2023 Pulsed laser-induced dewetting to fabricate Au-Pd nanoparticles supported on ZnO films and its application for the photocatalytic degradation of indigo carmine *Thin Solid Films* **776** 139884

[20] Zhang D, Ma Z, Spasova M, Yelsukova A E, Lu S, Farle M, Wiedwald U and Gökce B 2017 Formation mechanism of laser-synthesized iron–manganese alloy nanoparticles, manganese oxide nanosheets and nanofibers *Part. Part. Syst. Char.* **34** 1600225

[21] Wang J Q, Xi C, Wang M, Shang L, Mao J, Dong C K, Liu H, Kulinich S A and Du X W 2020 Laser-generated grain boundaries in ruthenium nanoparticles for boosting oxygen evolution reaction *ACS Catal.* **10** 12575–81

[22] Coviello V, Reffatto C, Fawaz M W, Mahler B, Sollier A, Lukic B, Rack A, Amans D and Amendola V 2025 Time-resolved dynamics of laser ablation in liquid with gasevolving additives: toward molding the atomic structure of nonequilibrium nanoalloys *Adv. Sci.* **12** 2416035

[23] Dzienny P, Rerek T, Szczęsny R, Trzcinski M, Skowroński Ł and Antończak A 2023 Laser-induced alloy nanoparticles: Au–Sn thin film morphology influence on the dewetting proccss *Int. J. Adv. Manuf. Technol.* **129** 1665–76

[24] Sachan R, Yadavali S, Shirato N, Krishna H, Ramos V, Duscher G, Pennycook S J, Gangopadhyay A K, Garcia H and Kalyanaraman R 2012 Self-organized bimetallic Ag-Co nanoparticles with tunable localized surface plasmons showing high environmental stability and sensitivity *Nanotechnology* **23** 275604

[25] Sachan R, Malasi A, Ge J, Yadavali S, Krishna H, Gangopadhyay A, Garcia H, Duscher G and Kalyanaraman R 2014 Ferroplasmons: intense localized surface plasmons in metal-ferromagnetic nanoparticles *ACS Nano* **8** 9790–8

[26] Ge J, Malasi A, Passarelli N, Pérez L A, Coronado E A, Sachan R, Duscher G and Kalyanaraman R 2015 Ferroplasmons: novel plasmons in metal-ferromagnetic bimetallic nanostructures *Microsc. Microanal.* **21** 2381–2

[27] Passarelli N, Pérez L A and Coronado E A 2014 Plasmonic interactions: from molecular plasmonics and fano resonances to ferroplasmons *ACS Nano* **8** 9723–8

[28] McKeown J T, Wu Y, Fowlkes J D, Rack P D and Campbell G H 2015 Simultaneous in-situ synthesis and characterization of co@cu core-shell nanoparticle arrays *Adv. Mater.* **27** 1060–5

[29] Cen X, Thron A M and Van Benthem K 2017 In-situ study of the dewetting behavior of Au/Ni bilayer films supported by a SiO_2/Si substrate *Acta Mater.* **140** 149–56

[30] Yao Y *et al* 2018 Carbothermal shock synthesis of high-entropy-alloy nanoparticles *Science (1979)* **359** 1489–94

[31] Yao Y, Dong Q, Brozena A, Luo J, Miao J, Chi M, Anapolsky A and Hu L 2022 High-entropy nanoparticles: synthesis-structure-property relationships and data-driven discovery *Science (1979)* **376** 1–11

[32] Allen F I, Pekin T C, Persaud A, Rozeveld S J, Meyers G F, Ciston J, Ophus C and Minor A M 2021 Fast grain mapping with sub-nanometer resolution using 4D-STEM with grain classification by principal component analysis and non-negative matrix factorization *Microsc. Microanal.* **27** 794–803

[33] Ophus C 2019 Four-dimensional scanning transmission electron microscopy (4D-STEM): from scanning nanodiffraction to ptychography and beyond *Microsc. Microanal.* **25** 563–82

Chapter 6

Nanoparticle properties and applications

In this chapter, we discuss the properties of metallic nanoparticles that are being applied to a large spectrum of applications. These properties include optical/plasmonic, magnetic, and catalytic properties. The scope of this chapter has excluded the discussion on biomedical properties and applications due to limited studies on nanoparticles formed by the laser-based techniques.

6.1 Optical properties and plasmonics

Metallic nanoparticles (MNPs), particularly those composed of noble metals such as gold, silver, and copper, exhibit unique optical properties that are markedly distinct from their bulk counterparts. These properties arise primarily due to the phenomenon of localized surface plasmon resonance (LSPR), a collective oscillation of conduction band electrons excited by incident light at specific frequencies. LSPR refers to the collective oscillation of conduction electrons at the nanoparticle surface in response to incident electromagnetic radiation. When the frequency of incident light matches the natural frequency of these oscillations, a strong enhancement in local electromagnetic fields occurs, resulting in intense light absorption and scattering at specific wavelengths [1–3], as shown in figure 6.1. While the collective oscillation of surface electrons occurs on the surface of most of the metals, and is also referred to as surface plasmon polariton, the phenomenon of LSPR has specifically enthralled the science community due to the tunability over the visible light spectrum and higher sensitivity.

When the dimensions of a metallic nanoparticle are smaller than the wavelength of light, the oscillating electromagnetic field can drive coherent electron oscillations at the nanoparticle surface. This resonance condition leads to strong absorption and scattering of light, which are highly sensitive to several factors, including particle size, shape, composition, surrounding medium, and inter-particle spacing.

doi:10.1088/978-0-7503-6224-5ch6

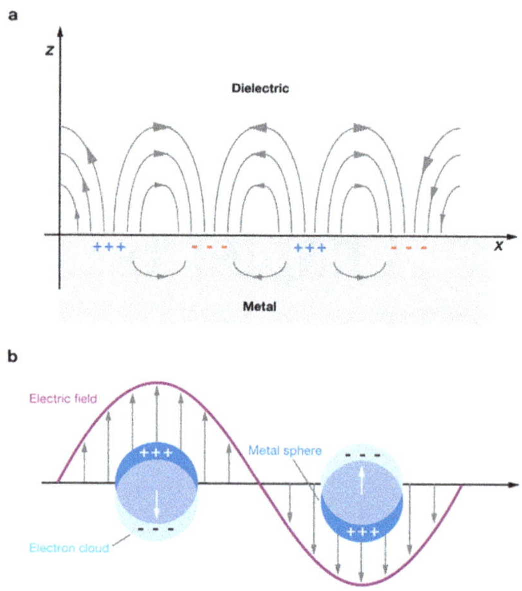

Figure 6.1. Schematic diagrams illustrating (a) a surface plasmon polariton (or propagating plasmon) and (b) a localized surface plasmon. Reproduced from [1], with permission from Springer Nature.

6.1.1 Surface plasmon resonance and size effects

The optical response of MNPs is fundamentally different from bulk metals due to confinement effects and increased surface-to-volume ratios. In bulk metals, the dielectric function follows the Drude model; however, at the nanoscale, size-dependent electron scattering and quantum confinement modify this behavior. For instance, gold and silver nanoparticles typically exhibit dipolar LSPR modes, with gold spheres showing a plasmon peak near 520 nm and silver near 400 nm [4, 5]. As particle size increases, radiation damping and retardation effects cause peak broadening and red-shifting [6].

For nanoparticles below ~10 nm in diameter, quantum size effects become prominent. Electron energy levels become discretized, the dielectric constant changes, and the classical Mie theory begins to break down [7, 8]. In this regime, nonlocal effects and surface scattering must be accounted for to accurately predict optical responses.

Various studies have been conducted showing the relationship between the particle size and LSPR characteristics, including the mono- and bimetallic nanoparticles synthesized using laser-based methods [9–16]. Wu *et al* demonstrated a size-dependent LSPR in pulsed laser-induced dewetting (PLiD)-synthesized pure Ag, pure Au, and Ag–Au alloy nanoparticles using electron energy-loss spectroscopy (EELS) [9]. Low-loss EELS characterization enabled by the monochromated electron beam captures the excitation of surface plasmons on the nanoparticle surfaces with high spatial and energy resolution [17–19]. As shown in figure 6.2, the nanoparticles showed red-shift in the LSPR energy position with increasing particle size [9]. Similarly, as the alloying

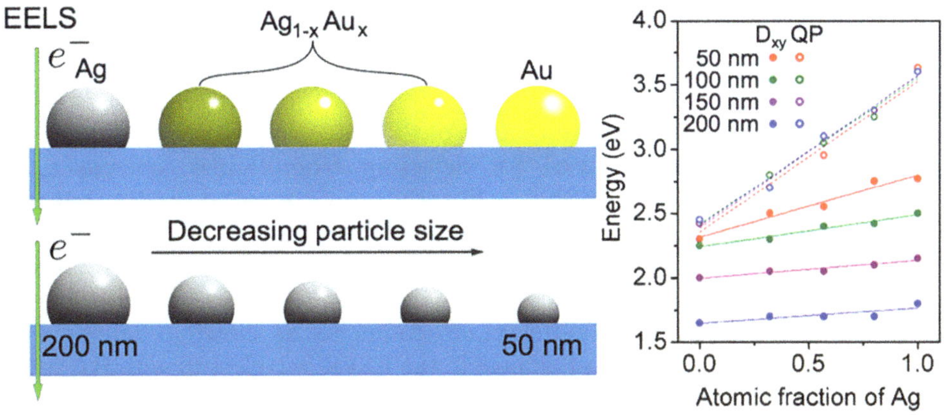

Figure 6.2. Schematic illustrating the electron energy-loss spectroscopic investigations on AgAu nanoparticles as a function of composition and particle size. The plot shows the measured quadripolar plasmon mode energy position with respect to changing size and composition. Reprinted with permission from [9]. Copyright (2016) American Chemical Society.

occurs between Ag and Au, a red shift in LSPR energy position is observed for nanoparticles of the same size. This shift is based on the effective domination of the dielectric function in the alloyed systems. It is also seen that the droplet-shaped nanoparticles obtained from pulsed laser dewetting show various modes of plasmon excitation due to asymmetry in the nanoparticles. In other studies conducted by Sachan *et al* [14] and Malasi *et al* [13], similar behavior of the presence of multiple modes has been reported in Ag–Co bimetallic nanoparticles, where Ag and Co make a thermodynamically immiscible system and exist as pure Ag and Co phase-separated regions within the nanoparticles.

6.1.2 Shape and morphology dependence

As mentioned briefly at the end of the previous chapter, nanoparticle shape plays a critical role in determining the plasmon resonance frequency and intensity. Spherical nanoparticles usually demonstrate a singular plasmon mode due to the symmetrical shape. On the other hand, anisotropic nanostructures such as nanorods, nanoprisms, and nanostars support multiple plasmon modes corresponding to different axes of electron oscillation. For example, gold nanorods exhibit two distinct plasmon bands: a transverse mode (short axis) in the visible region and a longitudinal mode (long axis) that can be tuned into the near-infrared by changing aspect ratio [20]. The sharp features in nanostars or prisms concentrate the electromagnetic field into ‘hot spots,’ which is beneficial for applications such as surface-enhanced Raman spectroscopy (SERS).

While there is not much variation seen in the geometry of nanoparticles formed by PLiD or laser ablation in liquids (LAL), a comparison of plasmonic properties with respect to the PLiD-driven intermediate and final stage nanostructures has been discussed by Yadavali *et al* [21].

6.1.3 Dielectric environment and refractive index sensitivity

The LSPR condition is extremely sensitive to the surrounding dielectric environment. A change in the refractive index (RI) near the nanoparticle surface leads to a measurable shift in the plasmon resonance peak. This sensitivity forms the basis of plasmonic sensing platforms used for detecting biomolecules, chemicals, and environmental changes [22]. The shift is linearly proportional to the RI change for small perturbations and can be further amplified using surface functionalization.

Although there is limited work reported on the sensitivity studies on laser-synthesized MNPs, a study by Sachan *et al* demonstrated the sensitivity of pure Ag and Ag–Co bimetallic MNPs in different environments, as shown in figure 6.3 [14]. Sensitivity is defined as the rate of change in the LSPR wavelength with RI, usually in units of nm/RIU. In figure 6.3, variation in the λ_{LSPR} as a function of the RI of the external dielectric environment is presented for various sizes and compositions of pure Ag and bimetallic Ag–Co nanoparticle arrays. The external dielectric was varied by encapsulating the samples in fluids with different refractive indices, as described in section 6.2. As evident from the plot, the LSPR shift is linearly dependent on the RI of the ambient environment. Ag–Co particle arrays show sensitivities of 51 and 71 nm/RIU, respectively, for 90 nm sized nanoparticles with compositions of 83.3% Ag and 71.5% Ag, whereas 110 nm sized Ag–Co particle arrays show sensitivities of 105 and 56 nm/RIU for the 83.3% Ag and 71.5% Ag compositions. These measurements indicated that the sensitivity of the Ag–Co bimetallic arrays, which ranged between 51 and 105 nm/RIU, was similar or comparable to pure Ag particles of similar shape and size, despite the bimetallic

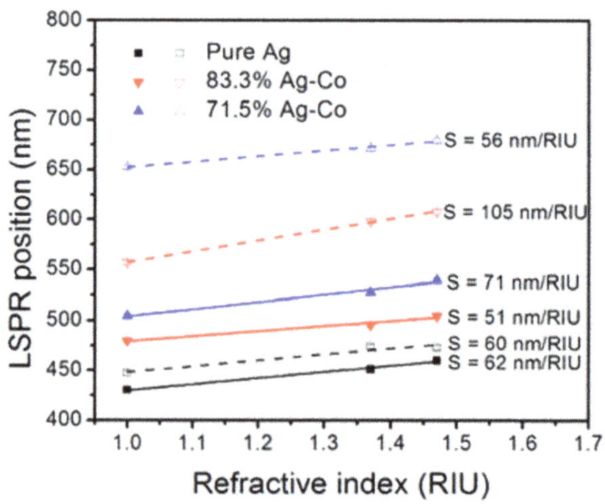

Figure 6.3. The shift in LSPR peak position as a function of RI of the external medium for various nanoparticle arrays. The plot shows the linear dependence of LSPR shift w.r.t. RI of external environment. Also, the slope depicts the sensitivity (in nm/RIU) of the respective nanoparticle arrays. The solid scatter points correspond to arrays with average particle size of 90 nm while open scatter points are for 110 nm. Reproduced from [14].

particles having a broader LSPR peak as compared to pure Ag. This study also demonstrated highly stable plasmonic characteristics in the nanoparticles occurring due to a galvanic coupling between Ag and Co.

6.1.4 Interparticle coupling and plasmon hybridization

When MNPs are in close proximity, their electromagnetic fields interact, leading to plasmon coupling and hybridization of individual plasmon modes. In nanoparticle dimers or chains, coupling can result in new bonding and antibonding plasmon modes with red-shifted absorption peaks [23, 24]. This coupling is strongly distance- and orientation-dependent and forms the basis of plasmonic waveguides and metasurfaces.

While the majority of the work on hybridization of plasmon modes and interactions between MNPs has been conducted on nanoparticles generated by chemical routes, a significant discovery of ferroplasmons is reported in Ag–CoFe bimetallic nanoparticles using the PLiD process, enabling nanoparticle formation with phase-separated Ag and CoFe regions. The study utilized monochromated EELS to probe the plasmon modes excited over Ag and CoFe regions with high spatial and energy resolution.

Typical experimental EELS spectra corresponding to marked locations on various nanoparticles under consideration (CoFe–Ag, Ag, and CoFe on carbon substrates) are shown in figure 6.4. Specifically, figure 6.4(b–d) shows high-angle

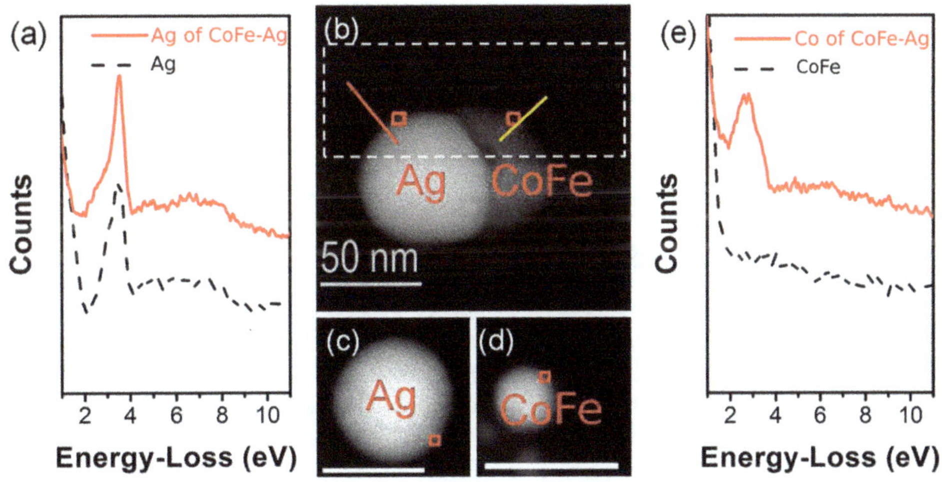

Figure 6.4. (a) Experimental EELS spectra from the surface of the Ag region in a CoFe Ag nanoparticle (solid line) and an isolated Ag nanoparticle (dashed line). High-angle annular dark field (HAADF) image of (b) a CoFe–Ag bimetallic NP, (c) an isolated Ag NP, and (d) an isolated CoFe NP. (e) Experimental EELS spectra from the surface of the CoFe region in a CoFe Ag nanoparticle (solid line) and an isolated CoFe nanoparticle (dashed line). The spectra were taken from the regions marked by square boxes in the respective nanoparticles. The scale bar for each HAADF image is 50 nm. All these nanoparticles were on C substrates. The nanoparticles shown in (b), (c), and (d) are of size 98 (along the major axis), 64, and 30 nm, respectively. Reprinted with permission from [16]. Copyright (2014) American Chemical Society.

annular dark field (HAADF) or Z-contrast images from the STEM of 98 nm CoFe–Ag, 64 nm Ag, and 30 nm CoFe nanoparticles, respectively. The report shows the observation of strong LSPRs in a nonplasmonic nanoparticle when in contact with another plasmonic material in the form of horizontally stacked nanoparticles on a substrate. These plasmons in metal-ferromagnetic nanostructures, or ferroplasmons for short, are in the visible spectrum with comparable intensity and bandwidth to those of the LSPRs from the Ag regions [16, 25]. This finding was enabled by electron energy-loss mapping across individual nanoparticles in a monochromated scanning transmission electron microscope. The appearance of the ferroplasmons is likely due to plasmonic interaction between the contacting Ag and Co nanoparticles. This study was followed by various efforts investigating bimetallic systems such as Ag–Ni, Ag–Au, Cu–Ni, and Ag–Al subsequently [11, 26, 27]. Moreover, a perspective article reported the potential mechanism of the ferroplasmon evolution through Fano resonance, which develops due to the interaction of two nanoparticles with different optical characteristics [28].

6.1.5 Applications leveraging optical properties

The tunable and strong optical responses of MNPs have led to a wide array of technological applications:

- **Biosensing:** LSPR sensors enable label-free, real-time detection of biomolecules [29].
- **Photothermal therapy:** MNPs absorb light and convert it to heat, enabling targeted cancer treatment [14].
- **SERS:** Enhanced local fields around MNPs dramatically increase the Raman signal of nearby molecules [30].
- **Photovoltaics and light harvesting:** MNPs enhance light absorption in solar cells via scattering and near-field effects [31, 32].
- **Metamaterials and optical devices:** Engineered nanoparticle arrays enable optical cloaking, negative refraction, and subwavelength imaging

6.2 Magnetic properties

MNPs synthesized using laser-based techniques, such as pulsed laser ablation in liquids (PLAL) or pulsed laser deposition (PLD), exhibit magnetic behaviors that differ significantly from their bulk counterparts. These deviations arise due to size effects, surface anisotropy, nonequilibrium crystallization, and chemical heterogeneity induced during rapid nucleation and growth under laser irradiation.

6.2.1 Surface and size-induced magnetic effects

At the nanoscale, the reduction in coordination number and quantum confinement effects lead to the emergence of magnetic moments, even in metals that are nonmagnetic in bulk. Gold, palladium, and platinum nanoparticles have demonstrated unexpected ferromagnetic behavior due to surface spin polarization and defect-related magnetism. These effects are particularly pronounced in particles synthesized via PLAL, where high-energy environments lead to defect-rich and metastable phases. The size-dependent ferromagnetism has been reported in Co nanoparticles, which show a transition from in-plane magnetization to out-of-plane as the particle size increases [33]. The nanoparticles below 75 nm predominantly demonstrate in-plane magnetization, whereas those above 75 nm show out-of-plane, as shown in figure 6.5. Interestingly, the nanoparticles up to 180 nm in size show single magnetic domain behavior. It is suggested that a size-dependent residual strain and the microstructure formed by rapid laser processing determine the orientation of nanomagnets [33].

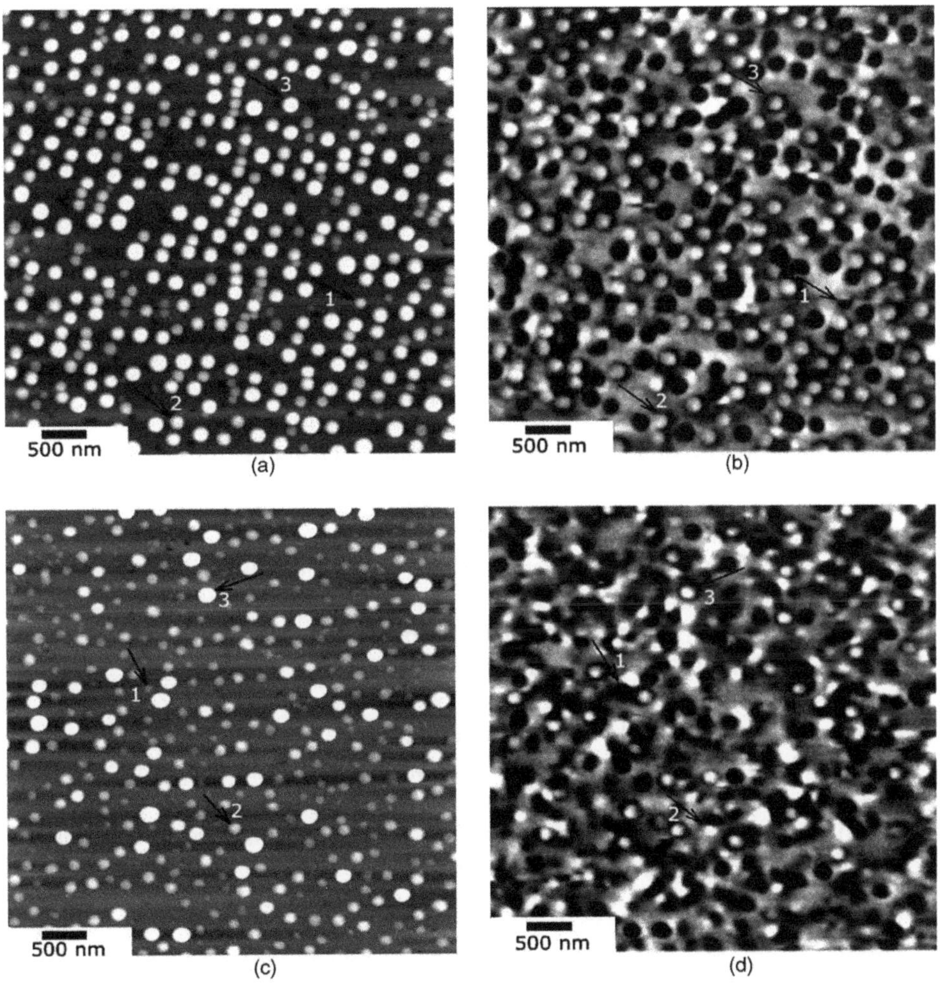

Figure 6.5. AFM and MFM images of as—prepared (a) and (b) and annealed (c) and (d) Co particles acquired under zero-field conditions. (a) and (b) nanoparticles identified by numbers 1, 2, and 3 represent in-plane, near 45°, and out-of-plane magnetization directions, respectively. (c) and (d) nanoparticles identified by numbers 1 and 2 represent in-plane, while number 3 represents out-of-plane magnetization direction. Reprinted from [33], with the permission of AIP Publishing.

6.2.2 Magnetic ordering and superparamagnetism

Iron, cobalt, and nickel nanoparticles synthesized using laser methods often exhibit superparamagnetism, where the magnetic domains fluctuate due to thermal energy. The superparamagnetic behavior depends on particle size, with a transition to ferromagnetism occurring above the blocking temperature. Tuning particle size, shell thickness (in core–shell structures), and oxidation levels allows for the control of coercivity and saturation magnetization [34].

6.2.3 Compositional effects and laser control

Laser-based synthesis allows for the precise formation of bimetallic and alloy nanoparticles such as FePt, CoPt, and FeCo, which are of particular interest for their high anisotropy and thermal stability. The formation of $L1_0$-ordered FePt phases, which are critical for magnetic data storage, can be facilitated by post-synthesis annealing or careful tuning of laser pulse parameters. Femtosecond lasers, compared to nanosecond pulses, often result in more monodisperse and crystalline nanoparticles with enhanced magnetic ordering [34].

6.2.4 Environmental and surface effects

The surrounding medium during laser synthesis plays a critical role in determining magnetic behavior. Water, ethanol, and ionic solutions can passivate surfaces, introduce strain, or promote oxidation, all of which affect magnetic properties. For example, Fe_3O_4 nanoparticles synthesized in ethanol may display higher magnetization due to limited oxidation compared to water-based synthesis.

6.2.5 Applications

These magnetic nanoparticles are utilized in biomedical imaging, targeted drug delivery, magnetic hyperthermia, high-density storage, and spintronic devices. Their behavior under alternating magnetic fields and their surface chemistry make them suitable for functionalization in biomedical applications.

Nanoparticles synthesized through laser-based methods such as PLiD, LAL, and PLD have emerged as highly tunable and active catalysts for chemical, electrochemical, and photochemical reactions. The unique thermal and kinetic conditions of laser synthesis produce nanostructures with specific features—such as clean surfaces, nonequilibrium phases, lattice defects, and tailored compositions—that are directly linked to catalytic performance.

6.3 Catalytic properties

6.3.1 Catalytic properties of nanoparticles from pulsed laser-induced dewetting

PLiD transforms a thin metal film into an array of nanoscale droplets via localized melting and retraction, followed by rapid solidification. The resulting nanoparticles are anchored to the substrate, often with uniform size and spacing. Catalytically, PLiD-derived nanoparticles offer (i) increased density of active surface sites due to high surface curvature and the absence of surfactants or capping agents; (ii) crystallographic control that allows exposure of catalytically favorable crystal facets; and (iii) alloy or core–shell configurations, especially when starting from multilayer films (e.g., Pt–Ni, Au–Pd), enabling synergistic catalytic effects.

These features have been exploited in reactions such as:

- CO oxidation, where Pt and Pd nanoparticles demonstrate high turnover due to defect-rich surfaces;
- Hydrogen evolution reactions (HER) and methanol oxidation, where the metal-support interaction and tailored morphology improve reaction kinetics.
- Crystal growth, where nanoparticles act as catalysts for vapor-phase growth of other materials.

One important case is the fabrication of Pt nanoparticles by PLiD for use as catalysts in semiconductor nanowire growth. Pt is a highly valued catalytic metal due to its role in electrocatalysis, photocatalysis, and nanowire synthesis. Conventional thermal dewetting of Pt films often requires high temperatures (500 °C–800 °C) and can result in incomplete dewetting or unwanted silicide formation when performed on Si substrates [35]. In contrast, PLiD of Pt thin films on SiO_2/Si substrates achieves rapid and complete dewetting at fluences of 322–400 mJ cm^{-2}, producing spherical, low-index-faceted nanoparticles without silicide contamination. The size of the Pt nanoparticles depends strongly on irradiation time at very short pulse durations, but stabilizes at ~10–20 nm for longer exposures, indicating a steady-state balance between laser heating and heat dissipation through the substrate. Interestingly, the nanoparticle size is largely independent of the initial Pt film thickness, a feature attributed to nucleation and hole growth dominating the dewetting dynamics. These PLiD-fabricated Pt nanoparticles have been demonstrated as effective catalysts for the chemical vapor deposition growth of silicon nanowires as shown in figure 6.6, following the vapor–liquid–solid (VLS) mechanism [35, 36]. The catalytic performance was influenced by the crystallographic

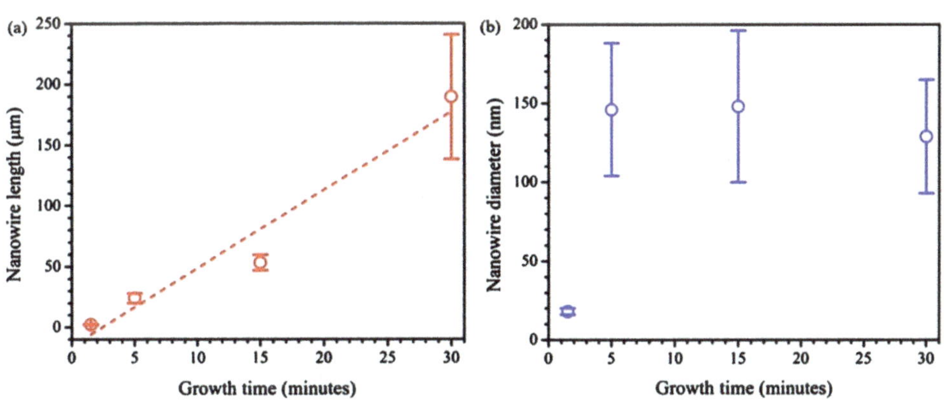

Figure 6.6. Plots of (a) SiNW length and (b) SiNW diameter versus the growth time at 950 °C using 14 ± 4 nm Pt nanoparticles from PLiD under the flow of 4 sccm SiCl4 and 100 sccm of H2. Reprinted with permission from [35]. Copyright (2025) American Chemical Society.

facets of the Pt nanoparticles, with spherical, low-index-faceted particles yielding slower nanowire growth than high-index-faceted ones, underlining the importance of nanoparticle morphology in catalysis.

Beyond monometallic systems, PLiD also enables the fabrication of bimetallic alloy nanoparticles with enhanced catalytic activity. For example, Au–Pd nanoparticles have been produced by irradiating sequentially deposited Au/Pd thin films supported on ZnO substrates [37]. A single ultraviolet laser pulse was sufficient to induce full dewetting and alloying, forming spherical nanoparticles of 50–110 nm diameter. X-ray diffraction confirmed the formation of solid-solution Au–Pd alloys, with composition tunable by adjusting the relative deposition thickness of the initial layers. These alloy nanoparticles exhibited significantly enhanced photocatalytic activity for the degradation of indigo carmine dye compared to monometallic Au or Pd nanoparticles, as well as bare ZnO films. In particular, Pd-rich compositions achieved near-complete dye degradation (99%), demonstrating the synergy of bimetallic catalysts where plasmonic response from Au combines with strong molecular adsorption from Pd [37]. The ability of PLiD to directly decorate semiconductor supports such as ZnO with alloy nanoparticles in a single step underscores its potential for photocatalytic and environmental applications.

Together, these studies highlight the versatility of PLiD in producing catalytically active nanoparticles. By tailoring parameters such as film thickness, laser fluence, irradiation time, and multilayer film composition, it is possible to control nanoparticle size, morphology, and alloy content. The direct, template-free, and surfactant-free nature of PLiD ensures that active metal surfaces remain exposed, maximizing catalytic efficiency. Applications demonstrated so far include semiconductor nanowire growth and photocatalytic pollutant degradation, but the method can be broadly extended to other catalytic reactions. Thus, PLiD provides a scalable and flexible platform for the design of advanced catalytic nanomaterials.

6.3.2 Catalytic properties of nanoparticles from LAL

LAL produces colloidal nanoparticles by ablating a solid target in a liquid medium. This method offers exceptional purity and tunability in nanoparticle composition and oxidation state, often without requiring reducing agents or stabilizers. The catalytic advantages of LAL-derived nanoparticles include [38]:

- Surfactant-free surfaces that enhance the availability of catalytic active sites;
- High density of undercoordinated atoms and surface defects, which act as catalytic hotspots or active sites;
- Tunability of oxidation states, especially for transition metals, which is critical in redox reactions.

Furthermore, pulsed laser in liquids synthesis of nanomaterials is rapid. Depending on the laser repetition rate, preparation of bulk quantities of tailored nanoparticles takes ca. 1 h or less. Scale-up to gram per hour productivities, which are relevant for industrial applications, has been demonstrated through reactor engineering [39, 40]. Protocol development for the synthesis of size and composition-controlled small

($\leqslant$ 50 nm) nanomaterials by conventional wet chemistry methods typically takes on the order of weeks to months. This long timescale impedes the preparation of systematic series or arrays of materials, in which only one property is varied at a time and which exhibit low polydispersity, but such nanomaterials are necessary to understand what controls chemical reactivity and catalytic cycles.

Laser-made nanocatalysts have been shown to be superior catalysts compared to analogues prepared by wet chemistry routes. For example, Muller and co-workers demonstrated that water oxidation electrocatalysts made by pulsed laser in liquids synthesis were intrinsically more active than analogous catalysts that were obtained from a wet chemistry method (figure 6.7) [41]. Laser-made Co_3O_4 nanoparticles outperformed commercial Co_3O_4 nanoparticles upon elimination of surface area differences between the two catalysts, suggesting a larger number of catalytically active sites in the laser-made material.

Another key advantage of pulsed laser in liquids synthesis of nanomaterials is its suitability for remotely controlled operation [42], enabling automatization, which is required for cost-saving industrial viability and real-world application of this powerful technique.

LAL-synthesized nanoparticles have shown superior performance in:

- Photocatalytic reactions and photoelectrocatalysis, using noble metal and transition metal oxide nanoparticles with enhanced light absorption and charge separation;
- Electrocatalysis, such as oxygen reduction reaction (ORR) and HER, where small Pt, Ag, or bimetallic nanoparticles demonstrate high mass activity;
- Environmental catalysis, including Fenton-like degradation and CO_2 reduction, enabled by Fe-, Cu-, or Mn-based oxide nanostructures.

Moreover, LAL can be extended to synthesize multimetallic or doped nanoparticles, which often outperform single-component catalysts due to electronic structure modulation [43, 44].

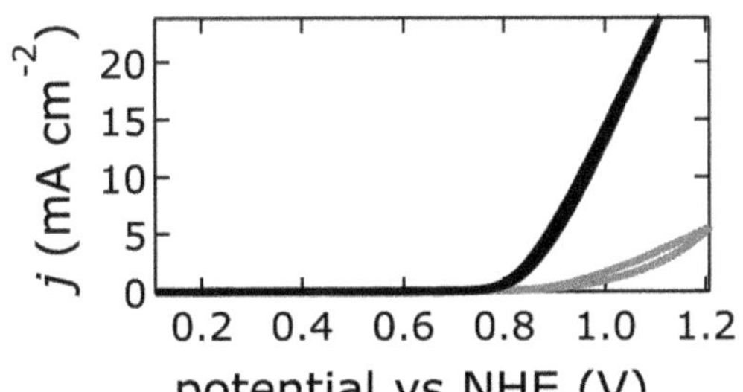

Figure 6.7. Laser-made nanocatalysts are intrinsically more active than analogous catalysts that were synthesized by conventional wet chemistry methods. Electrocatalytic performance of laser-made Co_3O_4 nanoparticles (black) was superior to that of commercial Co_3O_4 nanoparticles (gray); virtually identical catalyst masses were used. Current densities (*j*) were normalized to the calculated Co_3O_4 surface areas that were derived from measured particle sizes, effectively eliminating surface area differences between the two catalysts. Reprinted with permission from [41]. Copyright (2013) American Chemical Society.

As far as the photocatalytic performance of nanoparticles is concerned, photocatalysis and photoelectrocatalysis represent some of the most promising pathways for harnessing solar energy. More solar energy strikes the Earth in 90 minutes than humanity consumes in an entire year, making sunlight the largest, cleanest, and most ubiquitous energy source available. Harnessing this energy requires materials that can absorb light, generate and separate charge carriers, and efficiently drive surface redox reactions. Synthesis has emerged as a powerful platform to design such nanomaterials, enabling the fabrication of semiconducting and metallic nanostructures with structural and electronic features difficult to achieve by conventional methods [38].

The fundamental mechanism of light-driven catalysis involves three steps: (i) absorption of photons with energies above the semiconductor bandgap, generating electron–hole pairs, (ii) migration and separation of charge carriers to the surface or to cocatalysts, and (iii) participation of electrons and holes in surface redox reactions. LAL-synthesized nanoparticles, particularly semiconducting oxides, have demonstrated marked improvements in these processes due to the presence of laser-induced defects, lattice strain, and complex nanoarchitectures.

Bao and co-workers showed a striking example by generating nanocrystalline CoO via femtosecond laser irradiation (805 nm, 150 fs) of CoO micropowders suspended in water [45]. The laser-made CoO catalyzed overall water splitting with a stoichiometric H_2:O_2 ratio of ~2:1 and a solar-to-hydrogen efficiency of ~5%, while the starting micropowder was inactive. The enhanced performance was attributed to cobalt vacancies in the oxygen-rich nanocrystals, which shifted the flat-band potential by more than 1 V, facilitating charge separation and redox activity. Similarly, Shipley and co-workers used nanosecond UV laser irradiation (355 nm, 4 ns) of aqueous TiO_2 suspensions to generate yolk–shell-like microspheres composed of rutile shells and anatase-rich cores, decorated with disordered anatase nanoparticles [46]. Laser-induced oxygen vacancies imparted a blue coloration, lattice expansion, and defect-rich electronic states that boosted photocatalytic performance compared to pristine TiO_2. These examples underscore how ultrafast and nanosecond laser processes simultaneously control phase composition, defect concentration, and architecture in oxides, yielding photocatalysts with superior light-driven activity for solar energy conversion and environmental remediation.

In electrocatalysis, the role of nanostructuring is equally critical. Current densities scale with electrode area, so surface-area-normalized comparisons reveal intrinsic activity. LAL syntheses of advanced water oxidation electrocatalysts composed of the chemical elements, the majority of which are highlighted in figure 6.8, form not only mono- and bi-metallic nanoparticles but also various ceramic nanoparticles [38].

LAL-synthesized Co_3O_4 nanoparticles have emerged as highly effective oxygen evolution reaction (OER) catalysts. Müller and co-workers first reported quantum-confined Co_3O_4 nanoparticles (1.5–5.0 nm) generated by 355 nm, 8 ns pulses, achieving overpotentials as low as 314 mV at 0.5 mA cm^{-2} and turnover frequencies of 0.21 mol O_2 (mol Co, surface $s)^{-1}$ with 100% faradaic efficiency. Subsequent studies confirmed that oxygen-vacancy-rich cobalt oxides made by laser ablation of Co_3O_4 powders outperformed hydrothermal and RuO_2 benchmarks. Femtosecond

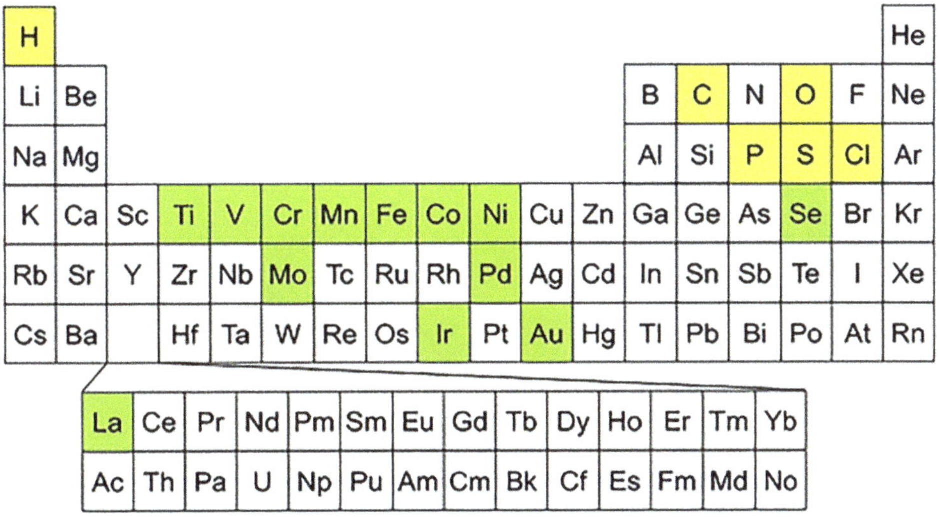

Figure 6.8. Chemical elements used for the pulsed laser in liquids synthesis of water oxidation electrocatalysts. Color code: metals and semiconductors (green), other elements (yellow). Reproduced from [38]. CC BY 4.0.

ablation of calcined CoOx produced ~5 nm multiphase nanoparticles (Co, CoO, CoOOH, Co_3O_4) that delivered mass activities exceeding 400 mA cm^{-2} mg^{-1}, with outstanding activity in both alkaline and neutral electrolytes. These results highlight the ability of LAL to produce defect-rich, multiphase nanostructures with exceptional electrocatalytic properties [41].

Beyond transition-metal oxides, LAL has been applied to noble metals and alloys, particularly for the ORR in fuel cells. Platinum nanoparticles synthesized by picosecond laser ablation in water (~3.8 nm) exhibited ~20% higher power density in proton exchange membrane fuel cells compared to commercial Pt/C, although durability challenges remained [47, 48]. The surfactant-free nature of these nanoparticles allowed facile surface functionalization, as demonstrated by Barcikowski and Yamamoto (figure 6.9), who decorated ~6 nm Pt nanoparticles with β-sheet peptides; the composites outperformed analogous citrate-stabilized materials made by wet chemistry, even if absolute activity was below commercial Pt/C [49].

A major advance came from Mukherjee and co-workers, who combined LAL with galvanic replacement chemistry to generate Pt–Co and ternary Pt–Cu–Co alloys [50]. Using 1064 nm, 4 ns pulses on a cobalt target in K_2PtCl_4 solutions, they synthesized ~4.7 nm Pt–Co alloy nanoparticles containing 22 at.% Co. These displayed excellent ORR activity in 0.1 M $HClO_4$, with mass and specific activities of 0.28 A mg^{-1} and 1.18 mA cm^{-2} at 0.9 V versus RHE, respectively, outperforming commercial Pt/C despite reduced Pt content. Expanding this method, $Pt_{72}Cu_{12}Co_{16}$ nanoalloys with a Pt–Cu–Co core and Pt-rich shell were produced, achieving performance superior to Pt/C while using ~28% less Pt. Notably, not all ternary alloys exhibited such synergy—Pt–Ru–Ni, Pt–Co–Mn, and Pt–Ni–Ti alloys were less effective—underscoring the importance of composition control in optimizing catalytic function.

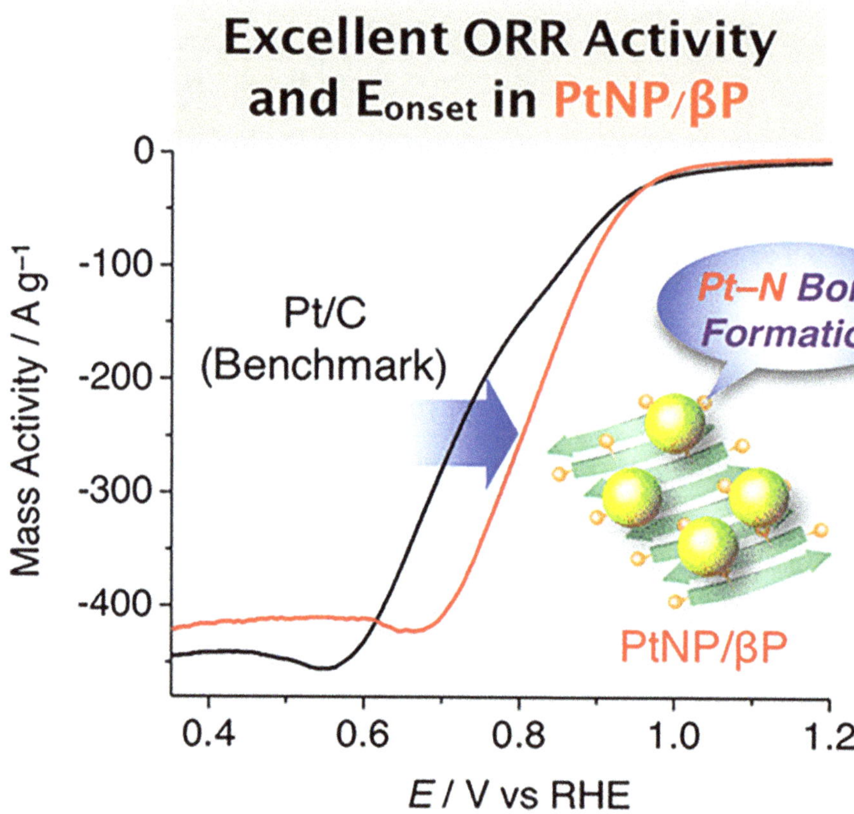

Figure 6.9. Oxygen reduction reaction (ORR) activities of laser-made Pt nanoparticles that were decorated with β-sheet peptides (red) and commercial Pt/C catalyst (black). Onset potential (E_{onset}) and mass activity of the laser-made material were ca. 100 mV inferior to the commercial benchmark system. Reprinted with permission from [49]. Copyright (2019) American Chemical Society.

Taken together, these examples demonstrate that LAL is not only a rapid and scalable synthesis method but also a platform for defect engineering, nano-architecturing, and compositional tuning of nanocatalysts. In photocatalysis, oxygen-vacancy-rich oxides like CoO and TiO_2 show enhanced solar water splitting performance. In electrocatalysis, vacancy-rich Co_3O_4 and carefully engineered Pt alloys deliver state-of-the-art OER and ORR activity, often surpassing conventional benchmarks. The ability to fabricate surfactant-free, compositionally tailored nano-catalysts rapidly and reproducibly positions LAL as a transformative tool for advancing clean energy and environmental technologies.

6.3.3 Catalytic properties of nanostructures from PLD

PLD's particle-by-particle growth (energetic plume, nonequilibrium surface kinetics) lets you (i) decouple particle size from loading, (ii) embed or partially embed metal nanoparticles in a stabilizing matrix in a single step, and (iii) control morphology via laser fluence, background gas, pressure, and plume flux. Those levers let you maximize under-coordinated surface atoms, create grain-boundary-rich crystallites, and engineer metal–support charge transfer—all three are recurrent motifs behind the biggest catalytic gains.

PLD-derived catalysts offer (i) high compositional fidelity, making it ideal for depositing complex oxides (e.g., perovskites, spinels) known for catalytic activity in OER and NOx reduction; (ii) tailored nanostructuring, which can increase the electrochemically active surface area; (iii) controlled oxidation environments, enabling phase-selective growth of catalysts such as Co_3O_4, $LaCoO_3$, and $SrIrO_3$.

While there is limited literature on metallic nanoparticle growth using PLD for catalytic applications, the process has been used for synthesizing metallic systems (e.g., Pt, Co), semiconductors (e.g., Si, Ge), and various oxide materials (e.g., ZnO, ITO). For example, catalyst-assisted PLD can guide growth through VLS mechanisms, yielding 1D nanowires and hierarchical assemblies. Morphological transitions, from nanowires at low background pressure to textured columnar films at higher pressures, illustrate how subtle parameter changes can strongly influence catalytic surface area and accessibility. These nanowires provide large surface areas and specific crystallographic orientations, both of which can tailor catalytic selectivity. Importantly, catalyst droplets (e.g., Au) can lower the growth temperature, allowing integration of catalytic structures on temperature-sensitive substrates [51].

Overall, the PLD-synthesized nanomaterials have shown remarkable activity in:

- **Oxygen evolution and reduction reactions**, where crystal phase, grain boundaries, and orientation strongly influence performance;
- **Fuel oxidation reactions** (e.g., ethanol and methanol oxidation), where Pt- or Pd-based thin films and nanoparticles exhibit stable, high-current responses;
- **Catalytic decomposition of environmental pollutants**, where doped oxide films enable high-temperature stability and sustained conversion.

Bibliography

[1] Maier S A 2007 *Plasmonics: Fundamentals and Applications* (Berlin: Springer)

[2] Kreibig U and Vollmer M 1995 *Optical Properties of Metal Clusters* (Berlin: Springer)

[3] Link S and El-Sayed M A 2003 Optical properties and ultrafast dynamics of metallic nanocrystals *Annu. Rev. Phys. Chem.* **54** 331–66

[4] Mulvaney P 2001 Not all that's gold does glitter *MRS Bull.* **26** 1009–14

[5] Amendola V and Meneghetti M 2009 Size evaluation of gold nanoparticles by UV–vis spectroscopy *J. Phys. Chem.* C **113** 4277–85

[6] Bohren C F and Huffman D R 1998 *Absorption and Scattering of Light by Small Particles* (New York: Wiley)

[7] Scholl J A, Koh A L and Dionne J A 2012 Quantum plasmon resonances of individual metallic nanoparticles *Nature* **483** 421–7

[8] Tame M S, McEnery K R, Özdemir K, Lee J, Maier S A and Kim M S 2013 Quantum plasmonics *Nat. Phys.* **9** 329–40

[9] Wu Y, Li G, Cherqui C, Bigelow N W, Thakkar N, Masiello D J, Camden J P and Rack P D 2016 Electron energy loss spectroscopy study of the full plasmonic spectrum of self-

assembled Au-Ag alloy nanoparticles: unraveling size, composition, and substrate effects *ACS Photonics* **3** 130–8

[10] Politano A and Chiarello G 2015 The influence of electron confinement, quantum size effects, and film morphology on the dispersion and the damping of plasmonic modes in Ag and Au thin films *Prog. Surf. Sci.* **90** 144–93

[11] Schletz D, Schultz J, Potapov P L, Steiner A M, Krehl J, König T A F, Mayer M, Lubk A and Fery A 2021 Exploiting combinatorics to investigate plasmonic properties in heterogeneous Ag-Au nanosphere chain assemblies *Adv. Opt. Mater.* **9** 2001983

[12] Collette R, Wu Y, Olafsson A, Camden J P and Rack P D 2018 Combinatorial thin film sputtering AuxAl1-*x* alloys: correlating composition and structure with optical properties *ACS Comb. Sci.* **20** 633–42

[13] Malasi A, Sachan R, Ramos V, Garcia H, Duscher G and Kalyanaraman R 2015 Localized surface plasmon sensing based investigation of nanoscale metal oxidation kinetics *Nanotechnology* **26** 205701

[14] Sachan R, Yadavali S, Shirato N, Krishna H, Ramos V, Duscher G, Pennycook S J, Gangopadhyay A K, Garcia H and Kalyanaraman R 2012 Self-organized bimetallic Ag-Co nanoparticles with tunable localized surface plasmons showing high environmental stability and sensitivity *Nanotechnology* **23** 275604

[15] Sachan R, Ramos V, Malasi A, Yadavali S, Bartley B, Garcia H, Duscher G and Kalyanaraman R 2013 Oxidation-resistant silver nanostructures for ultrastable plasmonic applications *Adv. Mater.* **25** 2045–50

[16] Sachan R, Malasi A, Ge J, Yadavali S, Krishna H, Gangopadhyay A, Garcia H, Duscher G and Kalyanaraman R 2014 Ferroplasmons: intense localized surface plasmons in metal-ferromagnetic nanoparticles *ACS Nano* **8** 9790–8

[17] Nelayah J, Kociak M, Stéphan O, García de Abajo F J, Tencé M, Henrard L, Taverna D, Pastoriza-Santos I, Liz-Marzán L M and Colliex C 2007 Mapping surface plasmons on a single metallic nanoparticle *Nat. Phys.* **3** 348–53

[18] Hachtel J A, Lupini A R and Idrobo J C 2018 Exploring the capabilities of monochromated electron energy loss spectroscopy in the infrared regime *Sci. Rep.* **8** 1–10

[19] Sachan R, Hachtel J A, Bhaumik A, Moatti A, Prater J, Idrobo J C and Narayan J 2019 Emergence of shallow energy levels in B-doped Q-carbon: a high-temperature superconductor *Acta Mater.* **174** 153–9

[20] Nikoobakht B and El-Sayed M A 2003 Preparation and growth mechanism of gold nanorods (NRs) using seed-mediated growth method *Chem. Mater.* **15** 1957–62

[21] Yadavali S, Khenner M and Kalyanaraman R 2013 Pulsed laser dewetting of Au films: experiments and modeling of nanoscale behavior *J. Mater. Res.* **28** 1715–23

[22] Anker J N, Hall W P, Lyandres O, Shah N C, Zhao J and Van Duyne R P 2008 Biosensing with plasmonic nanosensors *Nat. Mater.* **7** 442–53

[23] Prodan E and Nordlander P 2004 Plasmon hybridization in spherical nanoparticles *J. Chem. Phys.* **120** 5444–54

[24] Prodan E, Nordlander P and Halas N J 2003 A hybridization model for the plasmon response of complex nanostructures *Science (1979)* **302** 419–22

[25] Ge J, Malasi A, Passarelli N, Pérez L A, Coronado E A, Sachan R, Duscher G and Kalyanaraman R 2015 Ferroplasmons: novel plasmons in metal-ferromagnetic bimetallic nanostructures *Microsc. Microanal.* **21** 2381–2

[26] Wolf B S 2017 Dewetting properties of Ag-Ni alloy thin films *Master's Thesis* University of Tennessee
[27] Wu Y, Fowlkes J D and Rack P D 2011 The optical properties of Cu-Ni nanoparticles produced via pulsed laser dewetting of ultrathin films: the effect of nanoparticle size and composition on the plasmon response *J. Mater. Res.* **26** 277
[28] Passarelli N, Pérez L A and Coronado E A 2014 Plasmonic interactions: from molecular plasmonics and fano resonances to ferroplasmons *ACS Nano* **8** 9723–8
[29] Mayer K M and Hafner J H 2011 Localized surface plasmon resonance sensors *Chem. Rev.* **111** 3828–57
[30] Kneipp K, Kneipp H, Itzkan I, Dasari R R and Feld M S 2002 Surface-enhanced Raman scattering and biophysics *J. Phys. Condens. Matter.* **14** R597–624
[31] Atwater H A and Polman A 2010 Plasmonics for improved photovoltaic devices *Nat. Mater.* **9** 205–13
[32] Liu Y and Zhang X 2011 Metamaterials: a new frontier of science and technology *Chem. Soc. Rev.* **40** 2494–507
[33] Krishna H, Miller C, Longstreth-Spoor L, Nussinov Z, Gangopadhyay A K and Kalyanaraman R 2008 Unusual size-dependent magnetization in near hemispherical Co nanomagnets on SiO_2 from fast pulsed laser processing *J. Appl. Phys.* **103** 073902
[34] Chen M, Liu J P and Sun S 2004 One-step synthesis of FePt nanoparticles with tunable size *J. Am. Chem. Soc.* **126** 8394–5
[35] Ly L Q, Bonvicini S N and Shi Y 2025 Platinum nanoparticle formation by pulsed laser-induced dewetting and its application as catalyst in silicon nanowire growth *J. Phys. Chem.* C **129** 4553–64
[36] Schmidt V, Wittemann J V and Gösele U 2010 Growth, thermodynamics, and electrical properties of silicon nanowires *Chem. Rev.* **110** 361–88
[37] Segura-Zavala J, Depablos-Rivera O, García-Fernández T, Bizarro M, García-Morales R E and Sánchez-Aké C 2023 Pulsed laser-induced dewetting to fabricate Au-Pd nanoparticles supported on ZnO films and its application for the photocatalytic degradation of indigo carmine *Thin Solid Films* **776** 139884
[38] Forsythe R C, Cox C P, Wilsey M K and Mu A M 2021 Pulsed laser in liquids made nanomaterials for catalysis *Chem. Rev.* **121** 7568–637
[39] Sajti C L *et al* 2004 Generation of nanoparticles by laser ablation of metal targets in liquids: formation mechanism and particle size distribution *J. Phys. Chem.* B **108** 5507–11
[40] Kohsakowski S, Streubel R *et al* 2019 Controlling colloid formation by femtosecond laser ablation in liquids through the additive-mediated reaction environment *Phys. Chem. Chem. Phys.* **21** 6378–85
[41] Blakemore J D, Gray H B, Winkler J R and Müller A M 2013 Co_3O_4 nanoparticle water-oxidation catalysts made by pulsed-laser ablation in liquids *ACS Catal.* **3** 2497–500
[42] Crivellaro S, Guadagnini A, Arboleda D M, Schinca D and Amendola V 2019 A system for the synthesis of nanoparticles by laser ablation in liquid that is remotely controlled with PC or smartphone *Rev. Sci. Instrum.* **90** 033902
[43] Waag F, Li Y, Ziefuß A R, Bertin E, Kamp M, Duppel V, Marzun G, Kienle L, Barcikowski S and Gökce B 2019 Kinetically-controlled laser-synthesis of colloidal high-entropy alloy nanoparticles *RSC Adv.* **9** 18547–58

[44] Ghebretnsae S, Ziehl T J, Purdy S, Zhang P, Sham T K and Shi Y 2025 Pulsed laser-induced dewetting for the production of noble-metal high-entropy-alloy nanoparticles *Nanoscale* **17** 15423–35

[45] Liao L *et al* 2014 Efficient solar water-splitting using a nanocrystalline CoO photocatalyst *Nat. Nanotechnol.* **9** 69–73

[46] Balati A, Tek S, Nash K and Shipley H 2019 Nanoarchitecture of TiO_2 microspheres with expanded lattice interlayers and its heterojunction to the laser modified Black TiO_2 using pulsed laser ablation in liquid with improved photocatalytic performance under visible light irradiation *J. Colloid Interface Sci.* **541** 234–48

[47] Barcikowski S and Compagnini G 2013 Editorial: laser ablation in liquids: principles and applications in the preparation of nanomaterials *Phys. Chem. Chem. Phys.* **15** 3022–6

[48] Wagener P, Brandes G, Schwenke A and Barcikowski S 2011 Impact of in situpolymer coating on particle dispersion into solid laser-generated nanocomposites *Phys. Chem. Chem. Phys.* **13** 5120–6

[49] Jindal A, Tashiro K, Kotani H, Takei T, Reichenberger S, Marzun G, Barcikowski S, Kojima T and Yamamoto Y 2019 Excellent oxygen reduction reaction performance in self-assembled amyloid-β/platinum nanoparticle hybrids with effective platinum–nitrogen bond formation *ACS Appl. Energy Mater.* **2** 6536–41

[50] Hu S, Tian M, Ribeiro E L, Duscher G and Mukherjee D 2016 Tandem laser ablation synthesis in solution-galvanic replacement reaction (LASiS-GRR) for the production of PtCo nanoalloys as oxygen reduction electrocatalysts *J. Power Sources.* **306** 413–23

[51] Bazargan S and Leung KT 2012 Catalyst-assisted pulsed laser deposition of one-dimensional single-crystalline nanostructures of tin(IV) oxide: interplay of VS and VLS growth mechanisms at low temperature *J. Phys. Chem.* C **116** 5427–34

www.ingramcontent.com/pod-product-compliance
Ingram Content Group UK Ltd.
Pitfield, Milton Keynes, MK11 3LW, UK
UKHW051849100926
8556IPUK00005B/526

9 780750 362221